PRAISE FOR

Silent Earth

"A terrific book. . . . A thoughtful explanation of how the dramatic decline of insect species and numbers poses a dire threat to all life on earth."

—*Booklist* (starred review)

"Goulson's book is not only enormously informative, but also hugely entertaining: its light touch and constant humor make cutting-edge research a pleasure to read about. For anyone interested in the natural world, this is essential reading."

—*The Independent* (London)

"An attempt to educate us in the eccentric beauty of insects and their absolute necessity."

—*Sunday Times* (London)

"The book's greatest strength is its insistence that change is possible, and that everyone can make it happen in small and large ways. Goulson steps seamlessly between knowledgeable professor and impassioned environmentalist, and you can't help but get on board."

—*Sierra*

"An intense cri de coeur for saving insects. . . . The question now is whether they can survive what humans have done to the earth."

—*Times Literary Supplement* (London)

Silent Earth

Averting the Insect Apocalypse

DAVE GOULSON

HARPER PERENNIAL

NEW YORK • LONDON • TORONTO • SYDNEY • NEW DELHI • AUCKLAND

HARPER ● PERENNIAL

Originally published in the United Kingdom in 2021 by Jonathan Cape, an
imprint of Vintage.

A hardcover edition of this book was published in 2021 by HarperCollins
Publishers.

HarperCollins books may be purchased for educational, business, or
sales promotional use. For information, please email the Special Markets
Department at SPsales@harpercollins.com.

FIRST HARPER PERENNIAL EDITION PUBLISHED 2022.

Library of Congress Cataloging-in-Publication Data has been applied for.

ISBN 978-0-06-308821-4 (pbk.)

22 23 24 25 26 LSC 10 9 8 7 6 5 4 3 2 1

*For my crazy, frustrating, beautiful family and,
most of all, for my lovely wife Lara.*

Contents

Introduction: A Life with Insects   1

PART I: WHY INSECTS MATTER   7

1   A Brief History of Insects   9

2   The Importance of Insects   19

3   The Wonder of Insects   34

PART II: INSECT DECLINES   45

4   Evidence for Insect Declines   47

5   Shifting Baselines   68

PART III: CAUSES OF INSECT DECLINES   73

6   Losing Their Home   75

7   The Poisoned Land   85

8   Weed Control   120

9   The Green Desert   136

10 Pandora's Box   146

11 The Coming Storm   157

12 Bauble Earth   174

13 Invasions   179

14 The Known and Unknown Unknowns   189

15 Death by a Thousand Cuts   201

PART IV: WHERE ARE WE HEADED?   209

16 A View from the Future   211

PART V: WHAT CAN WE DO?   227

17 Raising Awareness   229

18 Greening our Cities   244

19 The Future of Farming   257

20 Nature Everywhere   278

21 Actions for Everyone   287

Acknowledgements   302

Further Reading   303

Index   318

Introduction

A Life with Insects

I have been fascinated by insects all my life. One of my earliest memories is of finding, at the age of five or six, some stripy yellow and black caterpillars feeding on weeds growing from the cracks in the tarmac at the edge of the school playground. I gathered them up, placing them among the crumbs in my empty lunchbox, and took them home. With the help of my parents I found more of the right type of leaves for them to feed on, and eventually the caterpillars transformed into handsome magenta and black moths (European readers might recognise them as cinnabar moths). This seemed like magic to me – and still does. I was hooked.

Since then, I've somehow managed to make a living from my childhood hobby. As a teenager I spent every weekend and holiday chasing after butterflies with a net, 'sugaring' for moths, and using pitfall traps to catch beetles. I bought the eggs of exotic moths from specialist mail-order suppliers and watched them grow into bizarre, rainbow-coloured caterpillars and eventually become huge, glorious moths: green moon moths with long tail streamers from India, peacock moths with flashing false eyes from Madagascar, and giant chocolate-brown Atlas moths, the biggest moth species of all, from South-East Asia. Inevitably I chose to study biology when I got a place at Oxford University, and later I did my PhD on butterfly ecology at Oxford Brookes, the rather less posh university perched on a hill to the east of Oxford. Following that, I managed to get various research posts: first, one back at Oxford University looking at the extraordinary mating habits of death-watch beetles, and then one at a government lab in Oxford studying ways to control moth pests by spraying crops with viruses. Since I didn't like killing insects I loathed this last job, and was enormously relieved when I was offered a permanent faculty position in the Biology Department at Southampton University.

It was there that I began specialising in bumblebees, for me the most endearing of insects (and they have stiff competition). I became fascinated by how bumblebees choose which flowers to visit, and spent five years unravelling how they avoid empty flowers by sniffing them for the faint whiff of a recent bumblebee visitor's smelly feet. I came to learn that, behind their bumbling, teddy-bear appearance, bumblebees are smart, the intellectual giants of the insect world, able to navigate and memorise the locations of landmarks and flower patches, efficiently extract rewards hidden in elaborate flowers, and live in complex social colonies where plots are hatched and regicide is common. Compared to them, the butterflies I had chased as a youth now seemed beautiful but airheaded creatures.

In pursuit of insects I have been lucky enough to travel the world, from the deserts of Patagonia to the icy peaks of Fjordland in New Zealand and the humid, forested mountains of Bhutan. I have watched clouds of birdwing butterflies sipping minerals from the muddy banks of a river in Borneo, and thousands of fireflies flashing their luminous bottoms in synchrony at night in the swamps of Thailand. At home in my garden in Sussex I have spent countless hours on my stomach, watching grasshoppers court a mate and see off rivals, earwigs tend their young, ants milk honeydew from aphids, and leaf-cutter bees snip leaves to line their nests.

I have had enormous fun. But I have been haunted by the knowledge that these creatures are in decline. It is fifty years since I first collected those caterpillars in the school playground, and every year that has passed since there have been slightly fewer butterflies, fewer bumblebees – fewer of almost all the myriad little beasts that make the world go round. These fascinating and beautiful creatures are disappearing, ant by ant, bee by bee, day by day. Estimates vary and are imprecise, but it seems likely that insects have declined in abundance by 75 per cent or more since I was five years old. The scientific evidence for this grows stronger every year, as more and more studies are published describing the collapse of monarch butterfly populations in North America, the demise of woodland and grassland insects in Germany, or the seemingly inexorable contraction of the ranges of bumblebees and hoverflies in the UK.

In 1963, two years before I was born, Rachel Carson warned us in her book *Silent Spring* that we were doing terrible damage to our planet. She would weep to see how much worse it has become. Insect-rich wildlife habitats such as hay meadows, marshes, heathland and tropical rainforests have been bulldozed, burnt or ploughed to destruction on a vast scale. The problems with pesticides and fertilisers she highlighted have become far more acute, with an estimated three million tons of pesticides now going into the global environment every year. Some of these new pesticides are thousands of times more toxic to insects than any that existed in Carson's day. Soils have been degraded, rivers choked with silt and polluted with chemicals. Climate change, a phenomenon unrecognised in her time, now threatens to further ravage our beleaguered planet. These changes have all happened in our lifetime, on our watch, and they continue to accelerate.

The decline of insects is terribly sad for those of us who love these little creatures and value them for themselves, but it also threatens human well-being, for we need insects to pollinate our crops, recycle dung, leaves and corpses, keep the soil healthy, control pests, and much, much more. Many larger animals such as birds, fish and frogs rely on insects for food. Wildflowers rely on them for pollination. As insects become more scarce, our world will slowly grind to a halt, for it cannot function without them. As Rachel Carson said, 'Man is a part of nature, and his war against nature is inevitably a war against himself.'

I now spend much of my time trying to persuade other people to love and care for insects, or at the very least to respect them for all the vital things they do. That, of course, is why I wrote this book. I want you to see insects as I do: as beautiful, surprising, sometimes surpassingly strange, sometimes sinister and disturbing, but always marvellous, and deserving of our esteem. You will, I think, be amazed at some of their more peculiar habits, life cycles and behaviours, which make the imaginings of science fiction writers seem ploddingly mundane. As we explore the world of insects, their evolutionary history, their importance, and the many threats that they face, the chapters are separated by short interludes – brief explorations of the lives of some of my favourite insects.

While time *is* running out, it is not yet too late to save the day. Our insects need your help. Most have not yet gone extinct, and if we just give them some space they can recover swiftly, for insects can breed fast. Insects live all around us: in our gardens, parks, farmland, in the soil beneath our feet and even in the cracks in a city pavement, so we can all get involved in looking after them, and ensuring that these vital creatures do not disappear. We may feel helpless in the face of many of the environmental issues that loom on our horizon, but we can all take simple steps to encourage insects.

I argue that we need profound change. We should invite more insects into our gardens and parks, turning our urban areas, and the interlinking road verges, railway cuttings and roundabouts, into a network of flower-filled, pesticide-free habitats. We need to radically change our broken food supply system, reducing food waste and meat consumption so that we can set aside huge areas of the less productive land for nature. We need to develop truly sustainable farming systems, focused on working with nature to produce food that is good for us, rather than growing commodity crops in vast, barren, pesticide- and fertiliser-soaked monocultures. We can all help drive these changes in many different ways: by buying and eating local, seasonal, organically-produced fruit and veg; by growing our own food; by voting for politicians who take the environment seriously; by educating our children on the urgent need to take better care of our planet.

Imagine a future in which our cities and towns are green, with every space filled with wildflowers, flowering and fruiting trees, green roofs and green walls; where children can grow up familiar with the chirp of grasshoppers, birdsong, the burr of passing bumblebees, the colourful flash of butterflies' wings. The cities are ringed by small, biodiverse farms producing healthy fruit and veg, pollinated by a diversity of wild insects, with pests kept in check by an army of natural enemies, and soil health and carbon stocks maintained by a myriad burrowing soil organisms. Further from the towns, new rewilding projects provide leisure opportunities for people to explore beaver-dammed wetlands brimming with dragonflies and hoverflies, flower-filled meadows and patchworks of woodland, all teeming with life. This may seem like a fantasy, but there is space on our planet for us all to live fulfilling lives, to eat healthily and well, and have a vibrant, green planet, brimming with life. We just have to learn to live as part of nature, not apart from it. And the first step is to start looking after the insects, the little creatures that make our shared world go round.

Part I

Why Insects Matter

I fear the majority of people don't much like insects. In fact, I would go further: I think many people loathe insects, or are terrified of them, or both. They are often referred to as 'creepy-crawlies', or 'bugs', the latter a term we also use for disease-causing organisms. For many of us, these terms are associated with unpleasant, scuttling, dirty creatures, living in filth and spreading disease. Increasingly, most of us live in cities, and grow up seeing few insects other than house flies, mosquitoes and cockroaches, so perhaps we should not be surprised that insects often inspire fear. Most of us are frightened of the unknown, of the unfamiliar. Few therefore appreciate how vitally important insects are to our own survival, and fewer still how beautiful, clever, fascinating, mysterious and wonderful insects are. My mission in life is to persuade people to love insects, or at least to respect them for all that they do. Here I want to explain why we should be teaching everyone from a young age to cherish these tiny creatures; why they matter.

A Brief History of Insects

Let's start at the beginning. Insects have been around for a very, very long time. Their ancestors evolved in the primordial ooze of the ocean floors, half a billion years ago, strange, armoured creatures with an external skeleton and jointed legs, known today to scientists as arthropods (meaning jointed feet). We have few fossils from that time, but those that exist, such as from the famous Burgess Shale deposits of the Canadian Rockies, give us tantalising glimpses of that early world. They are enormously diverse, with numerous types having body plans and numbers and shapes of limbs, eyes and other mysterious appendages unlike anything found today. It was as if Mother Nature had hit upon a successful concept, and was tinkering away like a child with a Meccano set, trying out different ways to bolt a creature together. For example, the aptly named *Hallucigenia* was a worm-like creature that was originally thought to walk on long, spine-like legs, with a crazy hairdo of waving tentacles on its back, but in more recent illustrations it has since been flipped over so that it walks on the tentacles and perhaps would have used the spines in defence. Meanwhile, *Opabinia* had five eyes on stalks, and a single lobster-like claw extending from its head, whereas *Leanchoilia* was a woodlouse-like creature equipped with two long arms at the front, each divided into three tentacles. Then there was *Anomalocaris*, an animal originally described as being three separate creatures – one shrimp-like, another a jellyfish, and the third something similar to a sea-cucumber – but it is now thought that these were all part of a single creature, the sea-cucumber being the body, the jellyfish its mouth parts, and the shrimp-like creature actually one of a pair of legs. At about 50cm long, *Anomalocaris* was the largest of the Burgess Shale fossils to have been described so far. We can only guess as to the behaviours and life cycles of these diminutive sea monsters of 500 million years ago. The early seas became crowded with these weird and wonderful creatures, but all are now extinct, although some must have founded lineages that are still present in the seas today.

What we do know is that a few of these early arthropods eventually experimented with moving to the land, perhaps to escape competitors or predators, or in search of their prey.

Having an external skeleton proved handy on land: most small sea creatures such as jellyfish and sea slugs rely on the water for support, and simply flop about in a helpless mess if they are stranded by a retreating tide, but with a rigid skeleton the early arthropods could walk, and this they did, exploring ever further from the water. They went on to found the most successful dynasty of creatures ever to tread the Earth. To this day, they are easily the most successful group on land, if you measure them by number of species or number of individuals (and not by their ability to mess up the planet). They, of course, are the insects.

Starting perhaps 450 million years ago, various different lineages of arthropod had a crack at life on land. Early arachnids dragged themselves from the sea and went on to become spiders, scorpions, ticks and mites – perhaps not the most glamorous of creatures to our human eyes, but very successful in their way. Millipedes ambled slowly onto land and occupied shady, damp habitats, quietly nibbling on decaying organic matter in the soil and under logs and stones, where they remain peacefully ensconced to this day. There, the millipedes were pursued by their fiercely predatory and rather faster relatives, the centipedes, also denizens of the soil and other dark, damp places.

Creatures of the Burgess Shale, animals that lived in the sea 500 million years ago: These weird creatures include many early arthropods, ancestors of the insects: sponges *Vanuxia* (1), *Choia* (2), *Pirania* (3); brachiopods *Nisusia* (4); polychaetes *Burgessochaeta* (5); priapulid worms *Ottia* (6), *Louisella* (7); trilobites *Olenoides* (8); other arthropods *Sidneyia* (9), *Leanchoilia* (10), *Marella* (11), *Canadaspis* (12), *Molaria* (13), *Burgessia* (14), *Yohoia* (15), *Waptia* (16), *Aysheaia* (17); molluscs *Scenella* (18); echinoderms *Echmatocrinus* (19); chordates *Pikaia* (20); along with *Haplophrentis* (21), *Opabina* (22), lophophorate *Dinomischus* (23), proto-annelid *Wiwaxia* (24), and anomalocarid *Laggania cambria* (25). From Wikicommons https://commons.wikimedia.org/wiki/File:Burgess_community.gif

A few crustaceans (crabs, lobsters, shrimps and so on) had a go at terrestrial life, but most never really got the hang of it. This group remains enormously diverse and abundant in the oceans to this day, but its most successful terrestrial representative is the humble woodlouse, an endearing and important creature in its way but with no serious claim to global domination.

The early arthropod adventurers on land, like woodlice and millipedes today, were presumably confined to damp places, along the water's edge, in mud, under stones or in clumps of moss. Aquatic creatures tend to die of dehydration very quickly on land, especially small ones like most arthropods. To really explore the land, waterproofing is vital. Spiders got the hang of this, evolving a waxy cuticle that now enables them to live even in the most arid places; I have seen them sitting patiently in their delicate webs, constructed on a scraggy leafless bush in the middle of the Sahara Desert. However, it was the insects that truly mastered terrestrial life. Their precise origin remains mysterious: insects are thought to have evolved on land about 400 million years ago,* perhaps from an early crustacean, perhaps from a millipede, but more likely to have come from some other ancient arthropod group that did not survive to the present and has not yet been found in fossils.

How, though, do we define or identify an insect? The answer is that all insects share certain common characteristics that distinguish them from other arthropods. Their body is divided into three sections: a head, thorax and abdomen. Unlike any other arthropod group, insects have six legs, which are attached to the thorax. Like the spiders, insects developed a waterproofed cuticle, sealed with waxes and oils.

Equipped with this basic design, insects set out to conquer the land, but they would probably not have got far were it not for one further huge evolutionary leap that was to be the key to their global success. One early insect took to the skies. Some primitive flightless insects still survive to this day – silverfish are perhaps the best known (which is to say, not very well known at all). Those able to fly, on the other hand, became enormously successful.

Powered flight has only evolved four times, so far as we know, in the three and a half billion years since life began, and insects were the pioneers of life in the air, about 380 million years ago (followed by the pterosaurs 228 million years ago, birds about

---

* Something pretty close to modern humans appeared approximately one million years ago, so insects have been here roughly 400 times longer than we have. They were ancient when the first dinosaurs appeared (about 240 million years ago), and have survived four of the Earth's five mass extinction events so far, including the one that wiped out the dinosaurs.

150 million years ago, and bats about 60 million years ago). For 150 million years insects had the skies to themselves. It isn't clear how flight first evolved, but a popular theory is that wings were originally flap-like gills, as seen in mayfly nymphs today. To start with they may have simply facilitated gliding, but eventually they became motile, and the first powered flight began.

Being able to fly bestows significant advantages. Escaping from land-bound predators becomes easy, and finding food or a mate is greatly facilitated, for flying is much quicker than walking. Migration becomes possible, with some insects such as the monarch and painted lady butterflies eventually evolving to fly thousands of miles each year to avoid the cold of winter. Migration is not a viable option if you are a woodlouse or a millipede.

With their new-found superpower the flying insects proliferated in the Carboniferous period (359 to 299 million years ago), with many new insect groups appearing, including the weak-flying mantises, cockroaches and grasshoppers, and also more accomplished flyers such as mayflies and dragonflies.

While the insects were busy learning to fly, the plants were not resting on their laurels. They too had developed better waterproofing of their leaves, and in competition with one another for light had grown ever taller, creating forests of giant tree ferns (some of which were of course to become fossilised as coal when they sank into the boggy forest floor). Although by this time there were amphibians and the first lizards, life on land must have been very largely dominated by insects. The air at the time was richer in oxygen than today, and that may be one reason why some insects were able to grow larger than any present-day species. If one could travel back to those ancient forests, one might glimpse *Meganeura* soaring between the trees – huge, dragonfly-like insects with wingspans of over 70cm.

Although flying may have been the insects' most important innovation, they had a couple more tricks up their six sleeves. Firstly, just after the end of the Carboniferous, about 280 million years ago, an insect species somehow achieved metamorphosis, the remarkable ability to change from a grub-like immature stage (the larva) into an adult insect with an entirely different appearance; from a caterpillar into a butterfly, or from a maggot into a fly.

Metamorphosis is as magical as any frog-to-prince transformation in a fairy tale, except that it is real, and happens all the time all around us. Imagine you are a full-grown caterpillar. You digest your final meal of leaves, then spin yourself a silken pad to hold you tight to a stem. You then split out of your old skin, revealing a new, smooth brown skin beneath. You no longer have eyes, or limbs, or any external openings except tiny holes called spiracles to allow you to breathe. You are entirely helpless, and will remain so for weeks, perhaps months in some species. Inside your shiny pupal skin your body dissolves, the cells of your tissues and organs preprogrammed to die and disintegrate, until you are little more than a soup. A few clusters of embryonic cells remain, and these proliferate, growing entirely new organs and structures, building you a brand-new body. Once it is ready, and the time is right, you split open your pupal skin and underneath have grown another one, this time complete with large eyes, a long, coiled proboscis for drinking, and beautiful wings covered in iridescent scales that you must inflate by pumping blood into their veins before they harden.

There is much debate as to how this astonishing phenomenon came about, including one recent and somewhat bizarre theory which suggested that metamorphosis evolved via a freak successful mating between a flying, butterfly-like insect and a velvet worm (a caterpillar-like relative of the arthropods). A more plausible suggestion is that caterpillars came about via the premature emergence of an embryonic insect from its egg. However they did it, metamorphosis is a remarkable phenomenon, and the insects that have this ability have become the most successful of all: flies, beetles, butterflies and moths, and wasps, ants and bees.

On the face of it, it may not be obvious why being able to transform oneself from a maggot into a fly is such a useful skill, impressive though it is. It seems like an awful lot of effort, and anyone who has ever reared butterflies can attest that emergence from the pupa is a delicate and precarious manoeuvre which often goes wrong, not least when wings fail to expand correctly, leaving the poor insect crippled and doomed. One theory as to why metamorphosis is such a successful strategy is that it enables the immature stages and the adults to each specialise in different tasks, and to have a body designed for the purpose.* The larva is an eating machine, little more than a mouth and anus connected by a gut, which is pretty much all that a maggot is. It does not need to be able to move quickly or travel far, as its mother will have ensured that she laid her eggs somewhere where food was plentiful. Larvae tend to have only rudimentary senses, with poor eyesight and no antennae. The adults, on the other hand, are often short-lived and feed rather little, other than perhaps sipping on some nectar to fuel their activity.† Their main task is to find a mate, copulate and, in the case of females, lay eggs. In some species they may also migrate. The adults need to be mobile and have acute senses, able to travel to seek out a partner by sight or smell or sound, so they often have large eyes and large antennae. They may also be equipped with bright colours to impress a potential mate.

For comparison, consider the lot of insects that do not go through metamorphosis. The grasshoppers or cockroaches, for instance. An immature grasshopper or cockroach is essentially a miniature version of the adult, with small wing 'buds' in place of functioning wings. Unlike insects that undergo metamorphosis,

---

* Please note that I am not suggesting intelligent design by a supreme being. 'Design' is shorthand for the blind tinkering of evolution over millennia.

† Insects being so numerous and varied, there are always exceptions. While some adult moths have no mouth parts and live for just three or four days, some adult beetles can live for several years. The record for insect longevity is held by queen termites which can live for at least fifty years, and possibly much longer.

these young grasshoppers may have to compete for food with adult grasshoppers, something that is not a concern for a maggot or caterpillar. The body of a grasshopper is essentially a compromised design which has to be able to do everything: feeding, growing, dispersing, finding a mate, finding somewhere good to lay eggs. To be fair to grasshoppers, they succeed pretty well, as any farmer in Africa who has faced a swarm of hungry locusts can attest, but in terms of numbers of species they are outclassed by their metamorphosing cousins. There are about 20,000 known species of Orthoptera (grasshoppers and their kin), and 7,400 species of Blattodea (cockroaches). In contrast, insects undergoing metamorphosis include 125,000 species of Diptera (flies), 150,000 species of Hymenoptera (bees, ants and wasps), 180,000 species of Lepidoptera (butterflies and moths) and an astonishing 400,000 species of Coleoptera (beetles). Together, these four groups of insects comprise about 65 per cent of all known species on our planet.

Aside from flight and the magic of metamorphosis, the final trick insects pulled off during their evolution was the development of complex societies, in which teams of individuals work effectively almost as if they were a single 'superorganism'. Termites, wasps, bees and ants all adopt this strategy, living in a nest with one or a small number of queens laying more or less all the eggs, and daughter workers performing various specialist jobs, such as caring for the queen, looking after the young, defending the nest and so on. By specialising, each individual insect can become an expert in its particular task, and in some cases even has a specially adapted body, as in for example the huge-jawed soldier castes found in some ant nests, principally engaged in defending the nest against attack by large predators such as anteaters or aardvarks. The famous American biologist E. O. Wilson, a specialist in ants, once estimated that there are in the region of between one and ten quadrillion individual ants in the world (1,000,000,000,000,000 to 10,000,000,000,000,000). In some terrestrial ecosystems they may make up 25 per cent of the total animal biomass, and overall the weight of ants on our planet is similar to the total weight of humans, to a very rough approximation. Ants alone outnumber us by about one million to one. Until perhaps the last 200 years, an alien looking down on Earth at any time in the last 400 million years would have concluded that this was the kingdom of the insects.

'Femme Fatale' Fireflies

Fireflies, also known in some countries as glow-worms, are surely among the most magical of insects. They are not flies at all, but a group of beetles possessing luminous bottoms. Their lights are used to attract a mate; different species glow green, yellow, red or blue, some producing a steady glow while others flash in a pattern particular to the species. In the European glow-worm, for instance, the female emits a gentle, steady green glow, which attracts the males. In many other species the glow is emitted in short flashes while in flight, which in the darkness creates the effect of a streak of light to the human eye, leading to another common name for these insects: lightning bugs. Some fireflies of the USA and tropical Asia glow in synchrony, creating a spectacular light display as thousands of insects flash their bottoms in unison.

Fireflies are predatory, variously feeding on other insects, worms or snails, depending on the species. Some female fireflies have even evolved the trick of mimicking the flash of the female of another species, not to attract a mate but to lure dinner. Unfortunate amorous males who respond to her allure are promptly consumed, from which habit these females are sometimes known as 'femme fatale' fireflies.

The Importance of Insects

If all mankind were to disappear, the world would regenerate back to the rich state of equilibrium that existed 10,000 years ago. If insects were to vanish, the environment would collapse into chaos.

E. O. Wilson, American biologist

In the autumn of 2017, I found myself doing a live interview about insect declines for an Australian radio show. The presenter's first, cheerfully delivered question was, 'So, insects are disappearing. That's a good thing isn't it?' I'm pretty sure the question was tongue in cheek, but since I was 12,000 miles away on the other end of a telephone, it was hard to be certain. Whatever the motivation, this question actually reflects the view of many people, who see insects primarily as pests, annoyances, spreaders of disease, stinging, biting, vexatious and bothersome. Few people bemoan the present-day lack of squashed insects on their car windscreen. Most of us now live in cities (83 per cent of the UK population are urbanites, according to the World Bank, with the global figure 55 per cent and rising fast), and unless we actually go looking for insects in our parks and gardens, we are most likely to encounter those that invade our homes, including cockroaches, house flies and bluebottles, clothes moths and silverfish. These are all fascinating and marvellous creatures but, as with a good malt whisky, one has to invest time in becoming properly acquainted with them before their merits become truly apparent. For most of us, they are unwelcome house guests, to be evicted or killed as swiftly as possible. For a moment I was nonplussed by the Australian interviewer's question, and additionally distracted because I was standing in a urinal at the time and someone had just come in to use the facilities.

I should say that I do not normally do radio interviews from public conveniences, but on this occasion I'd been dining in a pub en route to give a talk the next day in the English town of Dorchester, when the urgent request had come through on my mobile. The pub was playing loud music, and it was pouring with rain outside, so the toilet seemed the quietest and driest option. I gathered my wits together as best I could, and then launched into a well-practised diatribe about the many vitally important roles played by insects. Doing interviews like this is always disconcerting, as one can't see the interviewer's expression, and therefore any sense of whether one's points are coming across clearly – but at least the man urinating in the corner was nodding encouragingly.

A lack of enthusiasm for insects is certainly not confined to Australian radio show hosts. Recently on national BBC radio the eminent British doctor and TV presenter Lord Winston was asked about the global decline of wildlife. His reply? 'There are quite a lot of insects we don't really need on the planet.' Why he was asked to comment on a subject on which he has no expertise is unclear, but in these strange times it seems common for the opinions of celebrities to be valued regardless of qualifications or experience. Nonetheless, his response typifies the attitude of many.

Ecologists and entomologists should be deeply concerned that we have done such a poor job of explaining the vital importance of insects to the general public. Insects make up the bulk of known species on our planet, so if we were to lose many of our insects then overall biodiversity would of course be significantly reduced. Moreover, given their diversity and abundance, it is inevitable that insects are intimately involved in all terrestrial and freshwater food chains and food webs. Caterpillars, aphids, caddisfly larvae and grasshoppers are herbivores, for instance, turning plant material into tasty insect protein that is far more easily digested by larger animals. Others, such as wasps, ground beetles and mantises occupy the next level in the food chain, as predators of the herbivores. All of them are prey for a multitude of birds, bats, spiders, reptiles, amphibians, small mammals and fish which would have little or nothing to eat if it weren't for insects. In their turn, the top predators such as sparrowhawks, herons and osprey that prey on the insectivorous starlings, frogs, shrews or salmon would themselves go hungry without insects.

The loss of insect life from the food chain would not just be catastrophic for wildlife. It would also have direct consequences for the human food supply. Most Europeans and North Americans are repulsed by the prospect of eating insects, which is odd, since we happily consume prawns (which are broadly similar, being segmented, and with an external skeleton). Our ancient ancestors would certainly have eaten insects. Globally, however, eating insects is the norm, and in some countries insects make up a significant proportion of the diet. Roughly 80 per cent of the world's population regularly consume insects, with the practice very common in South America, Africa and Asia, and among the indigenous peoples of Oceania. Approximately 2,000 different insect species are regularly eaten by people, including caterpillars, beetle grubs, ants, wasps, moth pupae, stink bugs, grasshoppers and crickets. To give just a few examples, it is estimated that 1,600 tons of mopanie worms (large, juicy caterpillars of a species of emperor moth) are sold for human consumption in South Africa each year, with many more privately collected and consumed. In neighbouring Botswana, the mopanie worm trade is worth $8 million each year. The caterpillars are commonly dried and eaten as crispy snacks, tinned for longer storage, or eaten fresh, fried with onions and tomatoes. Exports of tinned silkworm pupae from Thailand are worth an estimated $50 million. In Japan, tinned 'inago' (a type of grasshopper) is widely sold as a luxury food, while a favourite dish of the late Emperor Hirohito was boiled wasps with rice. In Mexico, white maguey worms (caterpillars of a large skipper butterfly) and *ahuahutle* (eggs of aquatic bugs, sometimes also known as 'Mexican caviar') have long been harvested in bulk from the wild and even exported to the USA and Europe. However, trade in these two insects has declined in recent years, as the skipper butterfly has become scarce through over-collecting, while the bugs have declined due to water pollution.

These are mostly examples of the consumption of insects gathered from the wild, but a strong argument can be made that we humans ought to farm more insects as an alternative to pigs, cows or chickens. Conventional livestock waste lots of the energy they consume on keeping their body warm, and consequently are quite inefficient at converting vegetable material into human food – cows much more so than chickens. For example, a cow puts on about one kilogram of edible body mass for every twenty-five kilograms of vegetable matter consumed. Being cold-blooded, insects are much more efficient: crickets, for example, can put on one kilogram of digestible body mass from consuming just 2.1 kilograms of vegetable matter, making them twelve times more efficient. Insects are also far more efficient than cows in other ways: per kilo of food produced for human consumption, cows require fifty-five times as much water and fourteen times as much space, compared to crickets. What is more, insects are a healthier source of animal protein, being high in essential amino acids and much lower in saturated fats than beef.

There are other advantages of insect food. For example, we are much less likely to catch a disease from eating insects – with which we share no known diseases – compared to vertebrates (think mad cow disease, chicken flu or COVID-19, the latter thought to have come from bats or perhaps pangolins used in Chinese medicine).

Unlike cows, most insects produce little or no methane,* a powerful greenhouse gas, and they grow much faster than mammals. Arguably, animal welfare issues are avoided, since many insects can be kept at high density without apparent hardship, and in any case the capacity of an insect to suffer is presumably lower than that of a cow (though I know people who would disagree).

The point is that if we wish to feed the ten to twelve billion who are projected to be living on our planet by 2050, then we should be taking the farming of insects seriously as a more sustainable option to conventional livestock. My only issue with eating insects is that, of all the ones I've tried, none was particularly enjoyable – aside from chocolate-coated ants, that is, in which instance I'm pretty sure it was the chocolate that I was enjoying. But I have only tried a few types, and will endeavour to keep an open mind should I get the chance to sample fried mopanie worms or Mexican caviar.

While insects are scarcely ever eaten directly in Western societies, we do regularly consume them at one step removed in the food chain. Freshwater fish such as trout and salmon feed heavily on insects, as do game birds like partridge, pheasant and turkey. In Japan, freshwater fish such as smelt and eel form a significant part of the human diet. These fish are primarily insectivores, and thus the human food supply is directly dependent on the presence of sufficient freshwater insects. This relationship was made clear in 1993 when one of the largest lakes in Japan, Lake Shinji, was polluted with neonicotinoid insecticides running off agricultural fields. Populations of invertebrate in the lake fell sharply, leading to a dramatic collapse of the local fishing industry and costing hundreds of jobs. The

---

* Termites may be the exception. Termites are rather similar to miniature six-legged cows, having a special gut chamber filled with microbes that help them to digest cellulose and other tough plant materials. Just as the bacteria in the rumen of a cow produce methane, so do those inside termites, although scientists cannot yet agree how much, or whether their contributions to greenhouse emissions are something we should be worrying about.

average annual yields of smelt dropped from 240 tons between 1981 and 1992 to just 22 tons between 1993 and 2004, while the eel catch fell from 42 tons to 10.8 tons in the same period.

Aside from their role as food, insects perform a plethora of other vital services in ecosystems. Eighty-seven per cent of all plant species require animal pollination, most of it delivered by insects. That is pretty much all of them, aside from the grasses and conifers (which are pollinated by the wind). The colourful petals, scent and nectar of flowers evolved to attract pollinators. Without pollination, wildflowers would not set seed, and most would eventually disappear. There would be no cornflowers or poppies, foxgloves or forget-me-nots. We might lament our world slowly losing its colour, but an absence of pollinators would have a far more devastating ecological impact than just the loss of pretty flowers. For if the bulk of plant species could no longer set seed and died out, then every community on land would be profoundly altered and impoverished, given that plants are the basis of every food chain.

From a selfish human perspective, a loss of wildflowers might seem like the least of our worries, since approximately three-quarters of the crop types we grow also require pollination by insects. The importance of insects is often justified in terms of the ecosystem services they provide, which can be ascribed a monetary value, and pollination alone is estimated to be worth between $235 and $577 billion per year worldwide (these calculations aren't very accurate, hence the wide difference between the two figures). Financial aspects aside, we could not possibly feed the growing global human population without pollinators. We could produce enough calories to keep us all alive, since wind-pollinated crops such as wheat, barley, rice and maize comprise the bulk of our food, but living exclusively on a diet of bread, rice and porridge would quickly see us succumb to deficiencies of essential vitamins and minerals. Imagine a diet without strawberries, chilli peppers, apples, cucumbers, cherries, blackcurrants, pumpkins, tomatoes, coffee, raspberries, courgettes and runner beans, blueberries, to name just a few. The world already produces fewer fruit and vegetables than would be needed if everybody on the planet were to have a healthy diet (while also overproducing grains and oils). Without pollinators it would be impossible to produce anywhere near the 'five a day' fruit and veg we all need.

In addition to pollination, insects are important biocontrol agents (although this is a somewhat circular argument for the importance of insects, as many of the pests they control are also insects).* Nonetheless, were it not for predators such as ladybirds, ground beetles, earwigs, lacewings, wasps and hoverflies, among many others, pest problems on our crops would be much harder to manage, and we would be forced to apply many more pesticides. Without pollinators, we would have to rely more on those few wind-pollinated crops that do not need them, but this would then make it harder to rotate crops from year to year, in turn making pest problems even worse.

The role of insects in pest control is unglamorous, sometimes gruesome, and usually unappreciated. Wasps, for example, would rank pretty low on any poll of people's favourite insect, but that is perhaps because most people are unaware that the large majority of wasp species are parasitoids, many of which are extremely effective at reducing numbers of pests.† In my own garden my brassica crops – cabbages, broccoli, cauliflower and so on – are often attacked by the voracious caterpillars of both large and small white butterflies, which chew holes in the leaves and, unchecked, can reduce a cabbage plant to nothing more than

---

* In the interests of balance I should point out that, although insects perform many vital roles, they also deliver quite a lot of 'ecosystem disservices' too. Many of them vector human or livestock diseases, are pests of crops, or are parasites of livestock. Termites do a valuable job of breaking down dead wood, for instance, but can also be serious pests, consuming timber-framed homes in warmer climates.

† The word wasp conjures up images of the yellow-and-black-striped social wasps, but most wasps are much smaller, often entirely black and roughly ant-sized. They include the world's smallest insect, a species of fairy wasp which is just 0.14mm long.

a stem supporting a network of the tougher, inedible leaf veins. Luckily for me the damage is usually limited by the arrival of *Cotesia glomerata*. These are ant-sized wasps, black with yellow legs, the females equipped with a sharp egg-laying tube with which they inject clusters of eggs into each unfortunate caterpillar. The resulting grubs consume the caterpillars from the inside, eventually bursting out *en masse* to spin a cluster of tiny yellow cocoons around the fresh cadaver of their host. Even the more familiar large yellow-and-black-striped wasps that plague our picnics in late summer are far more useful than is generally understood. They are both pollinators of wildflowers and voracious predators of crop pests such as aphids and caterpillars; perhaps we should not begrudge them tucking into a tiny fragment or two of our human food as well.

Insects can also be valuable in controlling unwanted or invasive plants, such as the prickly pear cactus in Australia. Prickly pears from the arid regions of the Americas were introduced to Australia in the 1900s to use as living fences for livestock. To my mind these are horrible plants, for they are smothered in sharp, barbed spines that are exceedingly painful and difficult to remove from flesh – I once fell into a prickly pear bush in Spain when trying to study paper wasps – so it seems an odd choice of hedging plant. In any case, the plant was not content growing in straight lines and quickly spread out of control, covering 40,000 square kilometres of Queensland in north-eastern Australia in an impenetrable, spiny thicket. In 1925 a drab little moth from South America, *Cactoblastis cactorum*, was introduced, and in no time at all had eaten its way through almost all of the cacti.

Insects are also intimately involved in the breakdown of organic matter such as fallen leaves, timber, corpses and animal faeces. This is vitally important work, for it recycles the nutrients, making them available once more for plant growth. Most decomposers are never noticed. For example, your garden soil – and particularly your compost heap, if you have one – almost certainly contains countless millions of springtails (Collembola).

These minute, primitive relatives of insects, often under 1mm long, are named for their clever trick of firing themselves high into the air to escape predators, using their *ferculae*. This is a spring-loaded prong which normally lies flat against the underside of the abdomen, but can, when emergency strikes, be used to catapult the animal as far as 100mm. This army of minuscule high-jumpers does an important job, nibbling on tiny fragments of organic matter and helping to break them up into even smaller pieces which are then further decomposed by bacteria, releasing the nutrients for plants to use. Springtails are a vital and neglected component of healthy soils. Some are surprisingly cute too, with the more rotund species resembling tiny, chubby sheep (with a bit of imagination).

Decomposers may rarely be noticed, but their absence can have profound consequences, as cattle farmers in Australia found out in the mid-twentieth century. In most of the world, cowpats are fought over by an army of insects, and as a result don't last long. Within seconds or at most minutes of their sloppy deposition on the grass, the first dung flies and dung beetles will appear, attracted by the alluring odour plume wafting on the breeze. Dung flies lay their eggs, which hatch rapidly into maggots that then consume the decaying, bacteria-rich organic matter. Dung flies can complete their entire life cycle in about three weeks. Some dung beetles have aquatic ancestors, and the adults swim through the liquid dung while it is fresh using their paddle-like legs. Many dung beetles lay their eggs in the dung pat, while others create burrows in the soil beneath it into which they sequester dung for their offspring. A few roll balls of dung several metres from the pat in the hope of escaping from the insect throng. Predatory rove and ground beetles arrive to eat the dung-feeders, and birds such as crows and hoopoes are attracted to probe for insect grubs. The burrowing of the numerous insects aerates and dries the pat, which eventually crumbles to nothing, its nutrients successfully recycled.

Aside from releasing nutrients, efficient disposal of dung by insects provides a second valuable service to farmers: it plays a major role in getting rid of gut parasites in livestock. The eggs of parasitic worms are passed out of an infected animal in dung, from where they can contaminate the grass and be ingested by another cow or sheep. By burying and consuming the dung, insects quickly dispose of those parasite eggs. Ironically, the parasite treatments now given to cattle make their dung toxic to insects, slowing the recycling of dung and exacerbating the very parasite problem they were supposed to treat.

By contrast, the problem the first cattle farmers in Australia faced back in the nineteenth century was that there were no native Australian insects able to cope with sloppy dung. Adapted to the arid conditions, Australian mammals – the marsupials such as kangaroos and wombats – produce faeces of a very different consistency to cows: hard, pellet-like droppings. Historically, the Australian dung beetles had adapted to feeding on this material, but were almost entirely incapable of coping with the dung produced by the imported cattle of the first European settlers. As a result, the cowpats took years to break down and began to accumulate, smothering the pastures and leaving less and less grass for the livestock. With each cow producing about a dozen pats per day, by the 1950s the area of Australia covered by cowpats was estimated to be increasing by 2,000 square kilometres per year.

The solution of importing dung beetles able to cope with cow dung was proposed in the 1960s by Dr George Bornemissza, a recent immigrant from Hungary, and so the Australian Dung Beetle Project was born. Bornemissza spent the next twenty years travelling the world searching for suitable species of dung beetle to introduce to Australia, focusing mainly on South Africa as the source due to its similar climate. Some previous deliberate introductions of non-native species to Australia had gone horribly wrong: for example, cane toads from South America, introduced to help control sugar cane pests, have themselves become a plague, proliferating to the point where there are now estimated to be about 200 million of them, eating everything except the pests they were intended to control. The newly introduced dung beetles, by contrast, were a roaring success. Overall, twenty-three different dung beetle species were introduced, selected particularly for the speed with which they could remove dung, and between them they are able to thrive in the various different climatic regions of Australia. Today, thanks to those beetles, cowpats in Australia magically disappear within as little as twenty-four hours.

Other insects, the undertakers of the natural world, are similarly efficient at disposing of dead bodies. With uncanny speed, flies such as bluebottles and greenbottles locate corpses within minutes of death, laying masses of eggs that hatch within hours into wriggling maggots that race to consume the carcass before other insects arrive. Their relatives, the flesh flies, have an edge in this race, as they give birth directly to maggots, skipping the egg stage entirely. Just as with dung, the flies compete with beetles, in this case the burying and carrion beetles, which tend to be slower to arrive but consume both the corpse and the developing maggots. Burying beetles drag the corpses of small animals underground, lay their eggs on them, and then remain to care for their offspring, guarding them against other burying beetles and also culling and eating some of their offspring if they judge that there are too many of them for the remaining food supply. The sequence of arrival of different insect species and their rates of development are sufficiently predictable for any particular environmental conditions even to be used by forensic entomologists to judge the approximate time of death of human corpses when the circumstances of death are suspicious.

On top of all this, burrowing, soil-dwelling insects help to aerate the soil. Ants disperse seeds, carrying them back to their nests to eat but often losing a few, which can then germinate. Silk moths give us silk, and honeybees give us honey. In total, the ecosystem services provided by insects are estimated to be worth at least $57 billion per year in the United States alone, although this is a pretty meaningless calculation since, as the esteemed biologist and all-round good guy E. O. Wilson once said, without them 'the environment would collapse into chaos,' and billions would starve. What price do we put on avoiding that?

It's clear that many insects perform vital roles, but for most insects we simply do not know what they do. We have not even got round to naming four-fifths of the perhaps five million insect species that are thought to exist, let alone studied the ecological roles they might perform. In recent years drug companies have begun 'bioprospecting' the almost infinite number of chemical compounds found in different insects, and have found many new ones with potential medical uses, including new anti-microbial compounds which might help us tackle antibiotic-resistant bacteria, and also anticoagulants, vasodilators, anaesthetics and antihistamines. Every insect species that goes extinct means a treasure trove of potential drugs has gone forever.

As the conservationist Aldo Leopold said, 'To keep every cog and wheel is the first precaution of intelligent tinkering.' We are nowhere near understanding the multitude of interactions that occur between the thousands of organisms that comprise most ecological communities, and so we cannot say which insects we 'need' and which ones we do not. Studies of crop pollination have found that most pollination tends to be done by a small number of species, but that pollination is more reliable and more resilient over time when more species are present. After all, the numbers of different insect species fluctuate naturally from one year to the next; some might cope better with a cold spring, or heavy rain, or a drought, so the species that does most of the pollinating in one year might not be the main pollinator the following year, or in ten years' time. Relying on only one pollinator such as the domestic honeybee is a dumb strategy, because if anything happens to it there is no back-up.* As the

---

* Dumb it may be, but it is the strategy adopted by many farmers in North America who pay for honeybees to be shipped in to pollinate their crops, because their farming methods have rendered populations of resident wild bees too small for them to be able to provide adequate pollination.

climate changes so pollinator communities will change, and species that seem unimportant today could be the dominant pollinators of tomorrow. The same arguments can be applied to any of the other jobs that insects do; the more different types of insect we can hang on to, the better the chances that those vital jobs will keep getting done into our uncertain future.

The American biologist Paul Ehrlich famously likened loss of species from an ecological community to randomly popping out rivets from the wing of an aeroplane. Remove one or two and the plane will probably be fine. Remove ten, or twenty, or fifty, and at some point that we are entirely unable to predict, there will be a catastrophic failure, and the plane will fall from the sky. Insects are the rivets that keep ecosystems functioning. How close we are to the edge is unclear. In a few places we've already gone over. In parts of south-western China there are almost no pollinators left, and farmers are forced to hand-pollinate their apples and pears, as otherwise their crops would fail. In Bengal I have seen farmers hand-pollinating squash plants, and reports are coming in of farmers in parts of Brazil resorting to hand-pollinating passion fruits. Additionally, many studies around the world, ranging from blueberries in Canada to cashew nuts in Brazil, and French beans in Kenya, have found that the yields of insect-pollinated crops are lower in intensively farmed areas because of inadequate numbers of pollinators, whereas the yields are better on farms close to areas of native forests or other wildlife-rich areas, which act as a source of pollinating insects. In the UK, a recent study on Gala and Cox apple production found that farmers are currently losing about £6 million in potential income because the quality of their fruit is being impaired by inadequate pollination. Clearly in many parts of the world we are already at or beyond the point where a deficit of pollinators is limiting crop production. And if our crops are struggling to attract enough pollinators, then it seems likely that wildflowers may be too. If wildflowers decline further because of inadequate pollination, then this means even less food for the remaining

The Honeypot Ant

Bees and some wasps collect nectar from flowers and store it as honey in special cells or pots made from mud, paper or wax. These stores are vital in providing food through lean periods of the year when little is in flower. In the arid deserts of Australia lives a species of ant that has hit upon a quite different solution to the problem of nectar storage. The honeypot ant, properly known as *Camponotus inflatus*, has a unique trick. Individual ants themselves provide the storage, consuming so much nectar that their abdomen becomes grotesquely inflated. They quickly become unable to move, but their siblings continue to feed them until their abdomens become so stretched as to be transparent. Groups of well-stocked honeypot ants hanging from the roof of an underground nest chamber resemble a bunch of ripe golden grapes, each willing to regurgitate some of its store of sweet honey to any hungry member of the colony. These stores are so valuable in the parched Australian landscape that they attract thieves both large and small. Other ant nests will send raiding parties to overpower the guards and steal the helpless living food stores, dragging the swollen ants back to their own nest. Those honeypot ants are also hugely prized by the indigenous people of Australia, who will dig down up to two metres through the baked soil to reach them. The swollen ants are simply eaten alive, a delicious burst of sweet nectar.

3

The Wonder of Insects

There are undoubtedly strong practical and economic arguments for conserving insect species that either are, or might one day prove to be, valuable to humans. However, this anthropocentric approach to conservation is perhaps missing the most compelling arguments for conserving biodiversity. After giving talks I am often asked, 'What is the point of species X?' – for X, insert slugs, mosquitoes, wasps or any other creature the questioner happens to dislike. In the past, I tried to answer by constructing an ecological justification for the existence of species X, via its various roles, ideally including something that is useful to humans. For example, with slugs I might point out that they are the favourite food of slow worms, and are also eaten by many birds and mammals such as hedgehogs, and other creatures of which we tend to be fond. Some types of slugs help organic matter to decompose, some are predators of other slugs, and so on. Similarly, when I used to live in Scotland I was often asked what 'purpose' midges served. Visit the Highlands in late summer and you will soon learn to loathe these tiny brown flies, barely visible to the naked eye. They may be minuscule, but swarms of these blood-sucking little fiends can make life very uncomfortable, to the extent that in 1872 Queen Victoria is said to have once fled a highland picnic after being 'half-devoured' by midges. She is not the only one: midges are estimated to cost the Scottish tourist industry about £268 million a year in lost visits. But even midges have their important roles: they may be tiny, with a wingspan of about 2mm, and weighing just one two-thousandth of a gram,[*] but with up to 250,000 of them emerging from a single square metre of bog, that equates to about 1.25 tons per hectare of food for many birds such as swallows and martins, and for our smaller bat species. There are an astonishing 650 species of midge just in the UK, of which only about 20 per cent bite. Very little is known about the role their larval stages play, and in fact the larval stages of many species have yet to be described. In the tropics, for instance, midges include the sole pollinator of the cacao tree, meaning that without midges we would have no chocolate, so at least some of them are vitally important.

Of late, I have tried to turn the question around. Why should the existence of slugs or midges need to be justified, either for what they do for us, or for what they do for the ecosystem? Does there have to be 'a point' to slugs?

Remember here Aldo Leopold's comment that 'The first rule of intelligent tinkering is to keep all the parts.' Despite what he said, however, there do exist some insects and other creatures that could go extinct without us feeling any impact, ecological or economic. The Saint Helena giant earwig has already done so, and none of us noticed. Until a few decades ago it lived in seabird colonies on the remote Atlantic island, but this splendid 8cm beast hasn't been seen alive since 1967. It is a fair guess that it was wiped out by introduced rodents. Whatever ecological role it once performed, there seem to have been no obvious ecological repercussions from its loss, at least so far as anyone has noticed. New Zealand's giant wetas – huge brown

---

[*] Midges are miniature miracles of evolution. To fly they must beat their wings 1,000 times per second, the fastest wingbeat in the animal kingdom. Only the female bites, following the plume of carbon dioxide released as we exhale, which drifts downwind and enables her to track us down even in the dark. Once landed, she gyrates her head and scissors her sharp, serrated jaws to drive them into our skin using an action similar to that of an electric jigsaw. She can consume and carry off twice her body weight in blood.

armoured crickets, among the heaviest insects in the world, that plod slowly around the wet native forests – could follow it to oblivion, largely for similar reasons, and it is highly unlikely that there would be adverse consequences, save the heartbreak of a few New Zealand entomologists. Similarly, wart biter crickets could disappear from their last few haunts in the South Downs near where I live, and large blue butterflies could go extinct from the south-west of England, and I'm pretty confident that no ecological catastrophe would unfold as a result.

Maybe Robert Winston was right? Maybe 'we don't really need' lots of insects? Perhaps we humans could survive in a world with minimal biodiversity? The heavily farmed parts of Kansas or Cambridgeshire are pretty close already. Soon we may well have the power to eradicate entire species at will – for example, gene drive technology* can exterminate lab populations of the mosquito *Anopheles gambiae*, offering the possibility that one day we might be able to use it to wipe the species out in the wild (thankfully, this technology has not yet reached field testing). If we gain that power, should we use it, and where will we stop? Theoretically, a single release could wipe out the target species across entire continents, so how will such technology be policed at international level? Who gets to decide what lives and what dies? What species would

---

* This ingenious but terrifying approach involves genetically modifying the genome of the mosquito by inserting a dud version of a gene that is necessary for fertility in female mosquitoes. Females with just one copy of the defective gene are able to reproduce, but if they have two copies they are infertile. Alongside the dud gene, scientists insert a 'gene drive', a mechanism which ensures that every offspring inherits the defective gene, so that it rapidly becomes more common in the population. As it increases, so an ever larger proportion of the population inherits two copies, and becomes infertile. In lab populations, after eight generations all females were infertile, and so the populations were wiped out. Theoretically, releasing one engineered mosquito (or any other 'undesirable' organism such as a rat or cockroach) could wipe out entire wild populations, even entire species. However, it is unclear whether it would actually work so well in the real world, since in large wild populations some individual mosquitoes are very likely to evolve resistance, just as they do to insecticides.

be in the firing line after mosquitoes? Would it be a type of slug, or cockroach, or wasp? When would we decide that we had wiped out enough?

Technology is also being applied in a very different way. Robotics engineers in several labs around the world are developing robotic bees to pollinate crops, the premise being that real bees are in decline, and we may soon need a replacement. Is this the future we would wish for our children: one in which they will never see a butterfly flying overhead, where there are no wildflowers, and where the sound of birdsong and the buzz of insects is replaced by the monotonous drone of robot pollinators?

For me, the economic value of insects is just a tool with which to bash politicians over the head. They only seem to value money, so I point out to them that insects contribute to the economy. But if I'm honest, their economic worth has nothing whatsoever to do with why I try to champion their cause. I do it because I think they are wonderful. The sight of the first brimstone butterfly of the year, a flash of golden yellow wings in my garden on the first warm day in late winter, brings joy to my heart. Similarly, the chirrup of bush crickets on a summer's eve, or the sound of clumsy bumble-bees buzzing among the flowers, or the sight of a painted lady butterfly basking in the spring sunshine after her long migration from the Mediterranean – they all soothe my soul. I cannot imagine how desolate the world would be without them. These little marvels remind me what a wonderful and fascinating world we have inherited. Are we really willing to condemn our grandchildren to live in a world where such delights are denied them?

It is not just that insects can be beautiful. They are also fascinating, weird, utterly unlike ourselves in so many ways. Let me give just a few examples. Some tree hoppers – exotic relatives of aphids – have evolved to look exactly like a sharp thorn, presumably as camouflage but also making them tricky to swallow. Clusters of them feeding on the stem of a plant can make it appear to be a formidably prickly bramble. The ball-bearing tree hopper from Ecuador has a stalk emerging from the back of its head which branches and bears five hairy balls and a long backwards-pointing spike. It has been suggested that it is pretending to be infected with a *Cordyceps* fungus, which produces fruiting bodies that erupt from the head of their insect host, the theory being that predators might not like to eat infected prey. This has never been investigated. Nor has anyone researched why a large shield bug from Thailand bears an uncanny resemblance to Elvis. Some swallowtail caterpillars pull off a remarkably accurate impression of a bird dropping. Other caterpillars resemble spiders, flowers, snakes, twigs or seed pods. Lantern bugs, relatives of cicadas from the Americas, appear to be wearing a peanut shell on their head, for reasons unknown. The weevil family are most commonly rather drab, small, brown beetles, but males of the giraffe weevil from Madagascar are bright red and black, with a tiny head suspended on the end of an enormously long neck, which they use to try to dislodge rivals from the canopy in clumsy battles over females. Some male moths can extrude giant inflatable hairy appendages from their rear end, so called 'hair pencils' that help them to waft alluring pheromones onto the night's breeze.

Aside from an almost infinite variety of weird and wonderful appearances, insects have evolved a staggering variety of peculiar behaviours and life histories. For example, while most moths drink nectar, males of the vampire moth (*Calyptra* species) from Japan and Korea have a taste for blood, happily stabbing their serrated tongues into humans if the opportunity arises. Meanwhile in Madagascar there lives a moth, *Hemiceratoides hieroglyphica*, which sucks salty tears from underneath the eyelids of sleeping birds. In South America, scientists discovered sloth moths which as caterpillars feed only on sloth droppings, the adult moths hitchhiking among the fur of the sloth and laying their eggs on the fresh droppings as soon as they are produced.

Some of the many bizarre Central American tree hoppers described and illustrated by William Weekes Fowler (1894). See Further Reading.

The fruit fly *Drosophila bifurcata* produces sperm that are 5.8cm long, about twenty times longer than the diminutive fly itself. When inside the male they spend most of their life curled up as an impossible Gordian knot, but are somehow able to untangle themselves when inside the female. It seems that in this species larger sperm are somehow better at displacing smaller competitors in the race to fertilise the female's eggs.

In some cave-dwelling book lice from Brazil, it is the female that climbs onto the male for sex, inserting a large, inflatable and spiny penis-like structure into the male to suck up his sperm. The female's spiny penis helps keep them locked together until she has finished, which can take more than fifty hours. However, that is a brief tryst compared to some stick insects, the tantric sex masters of the insect world, which can remain in copulation for weeks, the record being seventy-nine days.

Even by insect standards, twisted wing flies are peculiar. They belong to their own obscure insect Order, the Strepsiptera, and very few people have ever seen one, although they are found all over the world, including in the UK. The female twisted wing fly lives as a parasite inside a bee, wasp or grasshopper, depending on the species. Once fully grown she may occupy 90 per cent of the space inside her unfortunate host, but the host somehow remains alive and active. Even as an adult the twisted wing fly female has no eyes, legs or wings, resembling a maggot, but this seemingly helpless creature pushes her blind head out between the segments of her host's abdomen, and releases a pheromone to attract a mate. The male is a small, delicate, free-flying insect with a single pair of dark, triangular wings. He mates with the female while she is still inside her host, and then promptly dies from the exertion. She gives birth to numerous live offspring, which consume her and then crawl out of the host's body and seek a fresh host. In the case of the twisted-wing flies that attack bees, the parasite larvae sit in flowers to await the arrival of a suitable bee, and hitchhike on it back to its nest, where they burrow into its offspring, so completing the bizarre life cycle.

Any one of these fantastic beasts could easily occupy a lifetime of study, or at the very least a fun PhD project. Of the one million insects we have so far named, most have not been studied at all, so who knows what fascinating discoveries remain to be made with regard to their lives? Given that there are thought to be in the region of another four million species that we have not yet even named, there is no doubt that scientists will be kept happily occupied studying them for millennia, so long as the insects remain to be studied. Would the world not be less rich, less surprising, less wonderful, if these peculiar creatures did not exist?

So, one can argue that insects are important, practically and economically, and one can argue that they bring us joy, inspiration and wonder, but both arguments are ultimately selfish, for both focus on what insects do for us. There is a final line of reasoning for looking after insects and the rest of the life on our planet, big and small, and it is one that is not focused on human well-being. One can argue that all of the organisms on Earth have as much right to be here as we do. If you are of a religious bent, do you really think that God created all of this amazing life just so we could recklessly destroy it? Do you think He or She intended for coral reefs to be bleached and dead, littered with plastic trash? Does it seem plausible that He or She went to the trouble of creating five million species of insect so that we could drive many of them extinct without ever even registering their existence?

If on the other hand you are not a believer, and accept the scientific evidence that species evolved over billions of years rather than being created by a supernatural being with a beetle obsession,* then you must realise that we are just a particularly

---

* The British evolutionary biologist J.B.S. Haldane was once asked what his decades of evolutionary research had taught him about the nature of God. His reply, perhaps tongue in cheek, was, 'He must have an inordinate fondness for beetles.' He might have added that God must also be pretty keen on wasps and flies.

intelligent and destructive species of monkey, nothing more than one of the perhaps ten million species of animal and plant on Earth. In that view, nobody granted us dominion over the beasts; we have no God-given moral right to pillage, destroy and exterminate.

Religious or not, most humans agree that the rich and powerful should not be allowed to oppress or dispossess the poor and the powerless (though of course we *do* allow it to happen all the time). Similarly, in dozens of sci-fi movies from *The War of the Worlds* onwards, aliens more intelligent than ourselves arrive, decide that the human race is redundant, and set about wiping us out so they can plunder the Earth for their own ends, or build an interstellar bypass. Of course, in these films we see the aliens as the bad guys, and we root for the inferior humans who usually somehow triumph in the end despite the odds being stacked against them.

When will we realise the hypocrisy of our position? On our own planet *we* are the bad guys, thoughtlessly annihilating life of all kinds for our own convenience. We intuitively grasp that the aliens in the movie *Independence Day* have no right to take our planet; I wonder what goes through the mind of an orang-utan as it sees its forest home bulldozed to the ground? There should not have to be 'a point of slugs' for us to allow them their existence. Do we not have a moral duty to look after all our fellow travellers on planet Earth, beautiful or ugly, providing vital ecosystem services or utterly inconsequential, be they penguins, pandas, or silverfish?

The Bombardier Beetle

Insects have evolved many fascinating defences against predators. Some, including various mantises, moths and grasshoppers, have superb camouflage, some have huge false eyes that make them seem large and dangerous, while others use bright colours to advertise that they have sequestered poisons in their body.

Few defences are as dramatic or effective as that of the bombardier beetle. This medium-sized, innocuous looking, ground-dwelling beetle has a unique skill. Its bottom contains a reservoir filled with a mix of hydrogen peroxide and hydroquinones. When under attack, these chemicals are squirted from the reservoir into a thick-walled reaction chamber lined with catalysts which cause the two chemicals to react violently, producing a contained explosion that the insect directs out of its rear end, spraying near-boiling and toxic benzoquinones towards the unfortunate assailant with an audible 'pop'. Small predators such as other insects may be killed outright, while larger predators such as birds are likely to beat a hasty retreat. I have had my fingertips scorched when inadvertently picking up one of these creatures, and can vouch that the experience is startling. The young Charles Darwin, an ardent collector of beetles, once placed a beetle in his mouth because he had run out of empty pots to put them in. It is lucky for him that it was not a bombardier beetle.

Part II

Insect Declines

4

Evidence for Insect Declines

It is widely accepted that we are now living in the 'Anthropocene', a new geological epoch in which the Earth's ecosystems and climate are being fundamentally altered by the activities of humans. I loathe the term, but I can't deny that it's appropriate.

A feature of this new era in Earth's history is the accelerating decline of biodiversity: the loss of wild animals and plants, and of whole communities of organisms. Public perception of this loss is particularly focused on extinction events, especially those of large mammals such as the mountain gorilla and African elephant, or birds which have already gone extinct like the passenger pigeon or dodo. These large, charismatic creatures capture the hearts and imaginations of the public, and are widely used by conservation organisations as 'flagship species', to raise funds for conservation work. It is heartbreaking to see film footage of the last northern white rhinos (at the time of writing there are just two left, both female), or of lonesome George, the last Pinta Island tortoise, shuffling about in their pens awaiting extinction.

Eighty species of mammal and 182 species of bird are known to have been lost in the modern era, usually defined as post-1500. Of course, this time period excludes the wave of 'megafauna extinctions' that took place in the late Pleistocene when man first spread around the world 40,000 years ago, wiping out almost all the large mammals and flightless birds that once roamed the Earth. However, evidence has recently begun to emerge that global wildlife has been affected far more profoundly than these relatively modest figures for numbers of extinct species might suggest.

Most species may not yet have gone extinct, but it is becoming clear that wild animals are on average far less abundant than they once were. A recent landmark paper by the Israeli scientist Yinon Bar-On estimated that since the rise of human civilisation 10,000 years ago wild mammal biomass has fallen by 83 per cent. To put it another way, roughly five out of every six wild mammals have gone. The scale of human impact is also revealed by his jaw-dropping estimate that wild mammals now comprise a meagre 4 per cent of all mammalian biomass, with our livestock (mainly cows, pigs and sheep) comprising 60 per cent and we humans making up the remaining 36 per cent. It is hard to grasp, but if he is correct then all the world's 5,000 wild mammal species – the rats, elephants, rabbits, bears, lemmings, caribou, wildebeest, whales and so many more – when combined, tot up to just one fifteenth of the weight of our cattle and pigs, and wild mammals are outweighed nine to one by the collective mass of humanity. The same scientist also calculates that 70 per cent of global bird biomass is now comprised of domestic poultry. The Anthropocene is upon us.

Also released in 2018 was the World Wildlife Fund and Zoological Society of London's 'Living Planet Report', which estimated that between 1970 and 2014 the total population of the world's wild vertebrates (fish, amphibians, reptiles, mammals and birds) fell by 60 per cent.* Within living memory, within my lifetime (I was born in 1965), more than half of our vertebrate wildlife has been lost. We are doing almost nothing to slow this decline (and much to accelerate it), so one has to

---

* When separated by habitat, freshwater vertebrates have declined most, at 81 per cent, compared to 36 per cent for marine vertebrates and 35 per cent for terrestrial vertebrates.

wonder what will be left after another forty-four years have passed? What world will our children inherit?

Catastrophic though declines of wild vertebrates have been, another even more dramatic change has been quietly taking place. This one may have more profound implications for human well-being. The large majority of the world's known species are of course invertebrates, meaning they lack a backbone, and the invertebrates are dominated on land by the insects. Insects are far less well studied than vertebrates, and for the majority of the one million species that have so far been named we know essentially nothing about them; their biology, distribution and abundance are entirely unknown. Often all we have is a 'type specimen' on a pin in a museum, with a date and place of capture. In addition to the one million named types of insect, there are estimated to be at least another four million species that we have yet to discover.* Although we are decades away from cataloguing the staggering insect diversity on our planet, evidence has emerged that these creatures are fast disappearing.

In 2015, I was contacted by members of the Krefeld Society, a mixed group of amateur and professional entomologists who since the late 1980s had been trapping flying insects in Malaise traps on nature reserves scattered across Germany. Malaise traps, named after their inventor, the Swedish scientist and explorer René Malaise, are tent-like structures that passively

---

* Given the hundreds of scientific expeditions that have taken place over the years, penetrating even the most remote regions of the Earth, this figure may seem scarcely credible. Take a net to a tropical forest, swish it about, and you will probably have caught species that are new to science; that is the easy bit. The difficult part is knowing which of the insects in your net are the new ones. For any particular insect you care to pull out, it would take weeks or months of specialist study, staring down a microscope, before you could be sure that the specimen did not belong to one of the million species that have already been named. There are very few people with the specialist knowledge to conduct this sort of work, so at the present rate of progress it will be hundreds of years before we have come close to cataloguing the full diversity of insect life.

trap any flying insects unlucky enough to bump into them. The German entomologists had amassed insects from nearly 17,000 days of trapping across sixty-three sites and twenty-seven years, a total of 53kg of insects (poor things). They sent me their data to ask for my help in making sense of it and preparing it for publication in a scientific journal. When I looked at the numbers and plotted some simple graphs I became increasingly fascinated and concerned. In the twenty-six-year period from 1989 to 2016 the overall biomass (i.e. weight) of insects caught in their traps fell by 75 per cent. In midsummer, when in Europe we see the peak of insect activity, the decline was even more marked, at 82 per cent. I thought initially that there must have been some sort of mistake, because this seemed

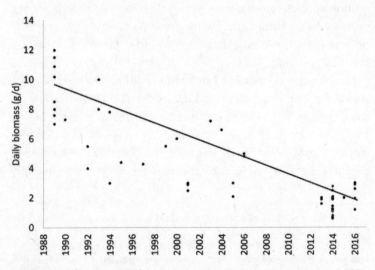

Declines in flying insect biomass on German nature reserves, 1989–2014: Insects were trapped using standardised Malaise traps (top right). The total weight of insects caught per trap per day declined by 76 per cent over the twenty-six years of the study. See Hallmann et al. (2017) in Further Reading.

too dramatic a drop to be credible. We knew that wildlife in general was in decline, but for three-quarters of insects to have disappeared so rapidly suggested a pace and scale of decline that had previously not been imagined.

We wrote up the work and tried to publish it in the most prestigious of scientific journals, *Nature* and *Science,* but they didn't think it was of sufficient interest. After some bouncing around the work was eventually published in a journal called *PLoS ONE.*

Thankfully, the study was then reported around the world, and has been much discussed since. Some argue that the data set is not robust, pointing out that it describes only biomass – weight, in another word – and that the insects had been neither identified nor counted. In a nutshell, what the critics were suggesting was that the loss in biomass could represent the disproportionate loss of just a few heavy insect species, and indeed it is theoretically possible that the actual *number* of insects might have remained stable or even increased, if big species were replaced by smaller ones. It was also pointed out that some of the sixty-three sites were sampled only in one year, while others were sampled multiple times over the duration of the study. In a perfect study with unlimited funding, one would have sampled every site in every year for the twenty-seven-year period. Ian Boyd, at the time the chief scientist of the Department for the Environment, Farming and Rural Affairs (DEFRA), the UK's government department responsible for the environment, was somewhat sceptical of the study, and pointed out that there might also be a subtle inherent bias in long-term population data sets such as this: scientists are more likely to start monitoring in places where they believe their creature of interest is common. Populations of all organisms tend to fluctuate over time; some will increase, others will fall. Larger-than-average populations are more likely to fall than to rise (a concept that statisticians call 'regression to the mean'). The phenomenon is perhaps most easily understood if one imagines the reverse situation. Suppose a scientist set up a network of monitoring sites in places where his chosen organism did *not* occur (which of course would be slightly crazy), then over time the population could only stay the same or rise. This said, the German study focused on nature reserves that remained intact and were managed to encourage wildlife throughout the study period. At the time, these German data were by far the best we had, and the pattern is very strong; it was hard to avoid the conclusion that there had been a major decline in insect biomass.

After publication of the Krefeld study in late 2017, a debate ensued as to whether similar declines in insect abundance were occurring elsewhere, or whether something peculiar was going on in German nature reserves. A partial answer came almost exactly two years later, in October 2019, when a different group of German scientists, led by Sebastian Seibold of the Technical University of Munich, published their findings from a very thorough study of insect populations in German forests and grasslands over ten years from 2008 to 2017. They studied 150 grassland sites and 140 forest sites, which encompassed various types of grassland across a spectrum from intensively farmed pastures to flower-rich meadows, and forest sites from managed conifer plantations to old broadleaf woodland. The grasslands were sampled using sweep nets to catch insects amongst the vegetation, while the forests were sampled with Perspex traps that caught mainly flying insects. Unlike the Krefeld study, Seibold and colleagues systematically gathered data from the same sites throughout the study, and the sites used included places that one would expect to be rich in insects alongside others that were likely to have few insects, thus overcoming the criticism raised by Ian Boyd. They also had the resources to count the more than one million arthropods they caught (including non-insects such as spiders and harvestmen), and to identify about 2,700 species. Given the short time scale – just ten years – the study's results were deeply troubling, for the rate of year-to-year decline was even greater than that described in the Krefeld data. Grasslands fared worst, losing on average two-thirds of their arthropod biomass (the insects, spiders, woodlice and more), one third of the species, and four-fifths of their total arthropod population. In woodlands, biomass dropped by 40 per cent, the number of species dropped by more than a third, and total arthropod abundance fell by 17 per cent (the latter not quite reaching 'statistical significance').* Yet, aside from a handful of applications of herbicides to the grasslands, none of these sites had been treated with pesticides. However, overall declines tended to be greatest in the sites with a higher proportion of surrounding agricultural land.

Taking the Krefeld study and the new data from Sebastian Seibold and colleagues, which collectively cover about 350 sites, it seems to be beyond reasonable doubt that insect populations in Germany have undergone dramatically rapid declines since at least the 1980s. As Professor William Kunin of Leeds University wrote in a commentary that accompanied the Seibold article, 'The verdict is clear. In Germany at least, insect declines are real, and they're every bit as severe as had been feared.'

What about elsewhere? Is there something peculiar going on in Germany? It seems highly unlikely. German land use and farming practices are pretty much the same as in neighbouring countries, governed largely by laws and policies that are common to the entire European Union. The countryside looks much the same as it does in, say, France, and the pesticides available are the same as elsewhere. So far as I can see, the only difference between Germany and the rest of us is that the Germans had

---

* Ecologists spend much of their time on often painfully complex analyses to try to discern whether patterns in their data could plausibly be due to chance. Somewhat arbitrarily, the accepted yardstick is a one-in-twenty chance. If there is a greater than one-in-twenty chance the pattern could arise by pure luck, then the pattern is deemed to not be significant. Conversely, if the probability is deemed to be less than one in twenty, the pattern is deemed to be likely to be real. In this case, all measures of decline were 'statistically significant' apart from the decline in total arthropod abundance in woodland.

the foresight to start monitoring their insects, while the rest of us did not, with the exception only of a few favoured insect groups. Hence hard data from elsewhere are largely lacking.

Only butterflies and moths have been monitored extensively and continuously elsewhere from 1970 onwards, in various localities from California via Ohio to Europe, and they show pervasive patterns of decline, though rarely as dramatic in magnitude as that found in Germany. The most high-profile example is the monarch (*Danaus plexippus*), a spectacular and iconic butterfly found across the USA and southern Canada in the spring and summer. It exists as two more or less discrete populations, one east of the Rockies and one to the west. The eastern monarch is famed for its long migration; in March, monarchs spread northwards from overwintering sites in the Sierra Madre Mountains of Mexico, breeding as they go, with successive generations reaching Canada by early summer. When autumn sets in they fly the 3,000 miles back from Canada to Mexico. The truly remarkable aspect of this migration is that the butterflies return to the exact same sites each autumn, even though it was their long-dead great grandparents that set out in the spring. How can they possibly know the way? Meanwhile, the western monarch undertakes a shorter but still impressive journey, from Canada to its overwintering sites in coastal California. In 1997, three forward-thinking Californian scientists, Mia Monroe, Dennis Fray and David Marriot, began counting the overwintering butterflies in their roosts, where they cluster together in trees to sit out the winter. It became an annual activity at Thanksgiving and New Year, with 200 volunteers now taking part, assisted and co-ordinated by the Xerces Society (a North American charity devoted to insect conservation). Sadly, Fray and Marriot both passed away in 2019, but they lived long enough to witness a terrible decline in their beloved monarchs. In 1997 the overwintering western butterflies in California had numbered some 1.2 million individuals, but in 2018 and 2019 there were fewer than 30,000, a drop of about 97 per cent. The eastern population has fared a little better, but in the ten years to 2016 numbers reaching Mexico fell by 80 per cent.

The monarch is not the only butterfly in decline. Perhaps the best-studied insect populations in the world are the UK's butterflies. They don't cluster conveniently together once a year to be counted, so instead they are recorded through the spring and summer by volunteers walking along transects as part of the Butterfly Monitoring Scheme. This scheme was set up by the forward-thinking entomologist Ernie Pollard, who worked for the (now defunct) Institute for Terrestrial Ecology, and was based at Monks Wood Research Station in Cambridgeshire, England. Pollard established a simple protocol in which a recorder walks along a fixed route every fortnight through the spring and summer, counting any butterfly within two metres either side of the path. Now often known as 'Pollard Walks', the protocol has since been borrowed around the world, and adapted for other insect groups. The scheme started out in 1976 with 134 transect sites, and today has more than 2,500, scattered across the UK. It is the largest and longest-running national insect recording scheme in the world. The trends it reveals are worrying. Butterflies of the 'wider countryside' – common species found in farmland, gardens and so on, such as meadow browns and peacocks – fell in abundance by 46 per cent between 1976 and 2017. Meanwhile habitat specialists, fussier species that tend to be much rarer, such as fritillaries and hairstreaks, fell by 77 per cent, despite concerted conservation efforts directed at many of them (although it should be noted that the first year included, 1976, was unusually hot in the UK and so was exceptionally good for these insects, accentuating the apparent decline).

Elsewhere in Europe, butterflies seem perhaps to be declining more slowly. For example, an analysis of trends in Europe-wide populations of seventeen grassland butterflies found a drop of 30 per cent between 1990 and 2011.

The butterflies' cousins, the moths, cannot be counted along transects because most of them are nocturnal. Many can, however, be attracted to lights and be lured into traps, and

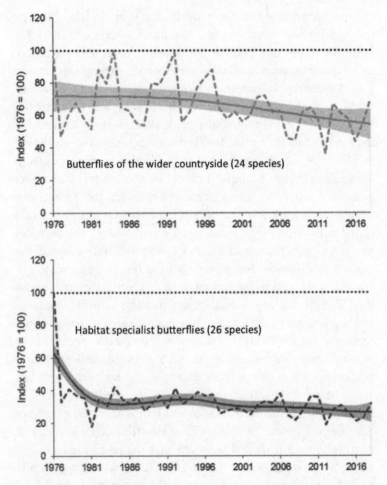

Trends in butterfly populations in the UK, 1976 to 2017: Numbers of butterflies recorded on transects across the UK vary from year to year, but broadly the pattern is one of decline. The upper diagram is for common, widespread species, which fell in abundance by 46 per cent; the lower diagram is for rare species, which fell by 77 per cent [Crown copyright, Department for Environment, Food and Rural Affairs, UK (2020). UK Biodiversity Indicators 2020].

light-trapping forms the basis of a long-running UK moth re-cording scheme. Moth trappers often focus on 'larger moths', which tend to be easier to identify, but as with their diurnal relatives all is not well. Overall abundance of larger moths in Britain fell by 28 per cent in the period from 1968 to 2007, with the decline more marked in the more urban and intensively farmed south of Britain, where the overall count fell by 40 per cent. A more recent analysis focusing on Scottish moths found a decline in abundance of 46 per cent in the period 1990 to 2014.

The only other large-scale and long-term UK data on insect populations come from suction traps designed to monitor aphid populations by sucking insects into the top of a 12 metre-tall tower. Aphids are weak flyers, but spread from crop to crop by flying upwards and then drifting on the wind like aerial plankton, which is why tall towers are used to sample them. The insects sucked into the top of the towers are sorted and counted by a small team of dedicated staff; they provide an early warning to farmers if an aphid invasion is detected. Data from four of these traps between 1973 and 2002 were ana-lysed by Chris Shortall at Rothamsted Research, Harpenden, the oldest agricultural research station in the world and the site of one of the suction traps. Although intended for aphids, the bulk of biomass in these traps was actually made up of one species of fly, the fever fly *Dilophus febrilis,* a weak-flying, black insect which presumably has the unfortunate habit of hanging about at 12m above the ground. Three of the four sites had rather few of any type of insect at the start of the study, and continued to have few throughout. The fourth site, in Herefordshire, started out with many more insects, but the biomass caught rapidly dropped by about 70 per cent over the thirty years of the study.

Since publication of the Krefeld research, scientists around the world have been searching for other long-term data sets, data from forgotten studies languishing unpublished in note-books or old Excel files. New publications are now appearing thick and fast, and almost all seem to show the same trend. In the Netherlands, for example, the biomass of ground beetles caught in pitfall traps fell 42 per cent in the period 1985–2017, while the biomass of moths in light traps fell by 61 per cent between 1997 and 2017. Also in the Netherlands, caddisflies – a group of moth-like insects with aquatic larvae – fell in abundance by about 60 per cent from 2006 to 2016, although the numbers of 'true bugs' (the common name given to insects belonging to the Hemiptera, which includes aphids, froghoppers and shield bugs) remained roughly stable. Meanwhile, across the Atlantic in coastal California between 1988 and 2018, meadow spittlebugs (another type of 'true bug') seem to have all but vanished. Aquatic insects in a river in Ghana fell by 45 per cent between 1970 and 2013. The data are still very patchy, but almost all the new evidence points in one direction: insects are declining, and fast.

You may be surprised that I have not yet mentioned bees. After all, the declines of bees have received much media attention due to their importance as pollinators. Unfortunately, however, there are no long-term data sets on the abundance of wild bee species. No one was organised enough to start trying to count them in a systematic way until recently. However, we do have accurate distribution maps for some of the better-studied wild bees, particularly bumblebees, obtained mainly from specimens in museum collections and from the small army of expert amateur recorders who, over many decades, have been keeping notes of the insects they see. One can use these to plot maps of the distributions of the various species at different periods in the past, and so see how the size of their geographic range is changing over time. For example, old records and specimens in museums show that the great yellow bumblebee (*Bombus distinguendus*) used to be found all over the UK, from Cornwall to Kent and north to Sutherland. Recent records are only from the far north and west of Scotland; it is extinct in England and Wales. We have no way of knowing how its UK population has changed (i.e. how many actual bees there are each year), but the geographic area it occupies has fallen by more than 95 per cent, so it seems certain that there are far fewer of them than there once were.

This approach reveals severe range contractions in many species. In the UK, geographic ranges of thirteen out of our twenty-three bumblebee species more than halved between pre-1960 and 2012, with two species (the short-haired bumblebee and Cullum's bumblebee) going extinct. Some caution is needed with these statistics, as the underlying records are mostly obtained by ad hoc recording by unpaid (though often very knowledgeable) amateur enthusiasts. The observed patterns depend a lot on how many recorders there are, how much time they can afford to spend on recording, where they happen to live or go on holiday, and so on (keen amateur entomologists may spend their entire holidays looking for insects, much to the irritation of their family). For example, if a fly enthusiast moved to, say, the English county of Lincolnshire, and spent his or her weekends searching for and recording flies, they would undoubtedly add lots of new records to the map (since not many people record flies). If they were subsequently to die or move away, it might seem to a scientist examining the data years later that there had been a period when flies flourished and then disappeared from Lincolnshire, while of course it is actually only the recorder that came and went.

The numbers of people recording insects, as well as the number of insect records obtained each year, have both greatly increased over time, which could make it seem as if particular species were becoming more widespread, or mask declines. Very recently, detailed analyses of patterns of range change of all Britain's wild bees (not just bumblebees), and also hoverflies, were carried out by Gary Powney of the Centre for Ecology & Hydrology in Oxfordshire, using complex mathematical techniques to try to take into account recorder effort. He found that both insect groups declined between 1980 and 2013, with an average of eleven species lost from each square kilometre of Britain. To put it simply, if you had searched any particular place in the UK for hoverflies and bees in 1980 and again in

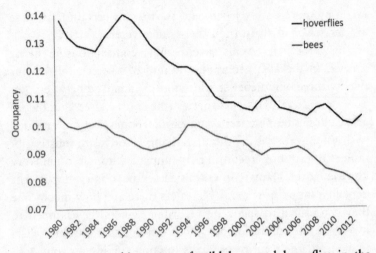

Changes in geographic ranges of wild bees and hoverflies in the UK: The trend lines show the average proportion of 1km grid cells occupied by each insect species in Britain. Wild bee species are shown in grey (based on 139 species) and hoverflies in black (based on 214 species). Thus, for example, in 1980, on average each hoverfly species occupied about 14 per cent of all 1km grid cells, but by 2013 it had fallen to about 11 per cent (from Powney et al., 2019).

2013, on average you would expect to find eleven fewer species on your second visit.

Twenty-three bee and flower-visiting wasp species have gone extinct in the UK since 1850. In North America, five bumblebee species have undergone massive declines in range and abundance in the last twenty-five years, with one, Franklin's bumblebee, going globally extinct. A more localised American study, of the bumblebees of Illinois, found that four bumblebee species went extinct in the state in the twentieth century. Meanwhile, in South America, the world's biggest bumblebee, *Bombus dahlbomii*, has gone from being widespread and common to close to extinction in just twenty years, caused by an invasion of disease-carrying European buff-tailed bumblebees. Even in the remote Tibetan plateau bumblebees appear to be in rapid decline, driven by overgrazing by domestic yak herds.

Moreover, although the bulk of insect species – the flies, beetles, grasshoppers, wasps, mayflies, froghoppers and so on – are not systematically monitored at all, we often have good data on population trends for birds that depend on insects for food, and these are mostly in decline. For example, populations of insectivorous birds that hunt their prey in the air (i.e. the flying insects that have decreased so much in biomass in Germany) have fallen by more than any other bird group in North America, by about 40 per cent between 1966 and 2013. Bank swallows, common nighthawks (nightjars), chimney swifts and barn swallows have all fallen in numbers by more than 70 per cent in the last twenty years.

In England, populations of the spotted flycatcher fell by 93 per cent between 1967 and 2016. Other once-common insectivores have suffered similarly, including the grey partridge (-92 per cent), nightingale (-93 per cent) and cuckoo (-77 per cent). The red-backed shrike, a specialist predator of large insects, went extinct in the UK in the 1990s. Overall, the British Trust for Ornithology estimates that the UK had 44 million fewer wild birds in 2012 compared to 1970.

All the evidence above relates to populations of insects and their predators in highly industrialised, developed countries. Information about insect populations in the tropics, where most insects live, is sparse. We can only guess what impacts deforestation of the Amazon, the Congo, or South-East Asian rainforests has had on insect life in those regions. We will never know how many species went extinct before we could discover them (most of the approximately four million species that we haven't named live in these forests). The American biologist Dan Janzen has been studying insects in Central America for an astonishing sixty-six years, and must know the insects of this region better than any other person alive, but to his regret he has not been monitoring insects in any systematic way. He is convinced that there have been massive declines. 'I have been watching the gradual and very visible decline of Mexican and Central American insect density and species richness since

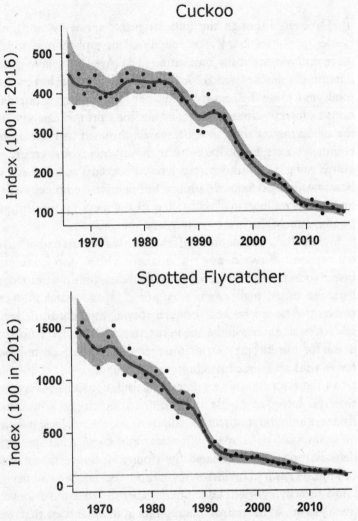

Population change of two insectivorous birds in England: The population index shown is scaled relative to 2012, which is set to 100. Thus one can see that the cuckoo population was just over four times larger in 1967 than in 2012 (top chart), while the spotted flycatcher population (bottom chart) was about fifteen times higher. Both species are specialists in eating insects, and both have undergone dramatic declines in England over the last fifty years. Within my memory, these have gone from being familiar, common birds to being such rarities that it is exciting to see or hear one. Reproduced from Massimino et al. (2020), with permission of British Trust for Ornithology.

1953,' he wrote recently. 'The house is burning. We do not need a thermometer. We need a fire hose.'

One long-term tropical study was recently published which confirms Janzen's view, providing perhaps the most concerning evidence of insect declines so far. In 1976 and 1977 the US entomologist Bradford C. Lister sampled arthropod abundance in the Luquillo Forest in Puerto Rico. He went there to study anolis lizards – a family of small, agile, insectivorous lizards with colourful, extendible throat pouches that they use to signal to one another and impress prospective mates. Lister was originally interested in whether there was competition for food between the different anolis species; at the time, working out how much competition occurs between species in nature was a hot topic. Since the anoles ate insects, he set about quantifying how many insects were present by using sweep nets and sticky traps. Returning to the same sites thirty-five years later, he repeated the sampling between 2011 and 2013. He found that the biomass of insects and spiders in sweep net samples had fallen between 75 and 88 per cent, depending on the time of year. Sticky trap sample catches had fallen by 97 to 98 per cent. The most extreme comparison was comparing identical sticky traps placed out in January 1977 versus January 2013, with the catch declining from 470mg of arthropods per day to just 8mg. 'We couldn't believe the first results,' Lister said in an interview. 'I remember [in the 1970s] butterflies everywhere after rain. On the first day back [in 2012], I saw hardly any.'

The Australian entomologist Francisco Sánchez-Bayo and colleague Kris Wyckhuys recently compiled all the long-term studies they could find relating to populations of wild insects – seventy-three in total. They found that there were huge knowledge gaps, with almost no data available from entire continents such as Africa, South America, Oceania and Asia, places hugely rich in insect life. The paucity of information on how our insects are faring, at a global scale, is also illustrated by the work of the International Union for the Conservation of Nature (IUCN). This body attempts to track and report the status of the Earth's wildlife with regard to extinction risk, highlighting species of particular concern in order to focus conservation efforts. The IUCN has assessed the status of every species of bird and mammal on Earth. In contrast, it has only been able to evaluate the status of just 0.8 per cent of the known insect species (likely to be less than 0.2 per cent of actual number of insect species). Sánchez-Bayo and Wyckhuys concluded that, although the long-term data on insect populations are spectacularly patchy, with no data at all for most insect groups and for many countries, the data we do have almost all point in the same direction: down. They concluded that, at a best estimate, insects are declining by about 2.5 per cent each year, with 41 per cent of insect species threatened with extinction. They also estimated that local extinctions of insects are happening eight times faster than in vertebrates, and stated that 'we are witnessing the largest extinction event on Earth since the late Permian' (the largest extinction event in Earth's history, which occurred 252 million years ago).

The Sánchez-Bayo and Wyckhuys study was criticised by some in the scientific community, who correctly pointed out that the authors had searched for studies using the key words 'insect' and 'decline', but had not also searched for 'increase', thus biasing the findings. Matters were made worse by a *Guardian* newspaper report on the article which extrapolated the 2.5 per cent decline per year and concluded that all insects may be extinct within a century – an unlikely claim since some insect species like house flies and cockroaches will certainly long outlast us humans.

A few months later, in early 2020, Roel van Klink and colleagues from a research centre in Leipzig published another global analysis, this time encompassing 166 long-term datasets on insect populations, and including studies that found increases. They concluded that, overall, terrestrial insects were declining at a rate of 9 per cent per decade, quite a bit slower than the Sánchez-Bayo and Wyckhuys study. Surprisingly, they found that freshwater insect populations had increased in recent years, driven in part by large rises in numbers of mosquitoes and midges at some sites. On the back of this, some even questioned whether insects were really in decline at all. However, this study too was then criticised, for it emerged that it suffered from a series of complex methodological flaws and errors, and inappropriately included data sets where human interventions had resulted in marked increases in local insect numbers. For example, one study was of dragonfly populations before and after creation of ponds for them to breed in; others were of insect populations in streams before and after clean-up operations to remove contamination with poisons. In such special circumstances it is hardly surprising that insect numbers increase, but this tells us nothing about global patterns. The exact rate of insect declines remains the subject of debate and, scientists being scientists, it is highly unlikely that we will ever agree.

A striking feature of the data we have on patterns of insect decline is that they only encompass very recent history, not even spanning my own lifetime (as I have mentioned, I was born in 1965). The earliest data we have are from the 1970s, with many studies such as those from Germany starting much later. The impact of mankind on the planet began long before 1989, the first year of the Krefeld study, which was twenty-seven years after the publication of Rachel Carson's *Silent Spring*, and more than forty years after the widespread adoption of synthetic pesticides. It seems probable that the 76 per cent drop in German insect biomass, if it is real, is just the tail end of a much larger fall. One recent study of butterflies in the Netherlands attempts to provide a window further back into the past, by analysing ranges of species represented in museum collections, similar to Gary Powney's analysis of UK bees and hoverflies, but going back to 1890. This approach suggests that the period of most rapid range contraction of butterflies in the Netherlands was between 1890 and 1980, well before the Krefeld study began. Overall, it estimates that butterfly ranges fell by 84 per cent over the entire 130-year period from 1890 to today. We will never know how many insects there were, say, a hundred years ago, before the advent of pesticides and industrial farming, but it seems certain that there were many times more than survive at present.

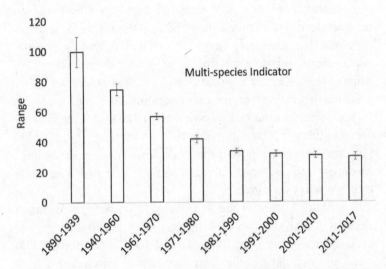

Patterns of range change for butterflies in the Netherlands, 1890–2017: The patterns are estimated from the locations of museum specimens, and are based on seventy-one species. The range changes are show relative to a value of 100 in the first time period. Declines appear to have been fastest in the first half of the twentieth century, before any detailed insect monitoring began (from van Strien et al., 2019).

The Emerald Cockroach Wasp

One of the most beautiful yet sinister of insects is the emerald cockroach wasp, a slender, 2cm-long metallic green creature with bright red legs, found in much of tropical Africa and Asia.

The female wasp seeks out her preferred prey, which are large cockroaches of the types commonly found infesting less salubrious houses and restaurants. Once she finds one, she pounces on it and swiftly stings the roach in the thorax, inducing temporary paralysis. With the cockroach now unable to move, she then carefully inserts her stinger into the precise part of the victim's brain that controls its escape reflex, and permanently disables this with a second venom injection. She then passes a little time by chewing off half of each of her victim's antennae, and drinking the drops of haemolymph (insect blood) that exude, allowing her first sting to wear off and the second to take full effect. The cockroach becomes docile, almost zombie-like, and although it is much larger than the wasp she is able to grasp the stump of an antenna in her jaws and lead it back to her burrow like a dog on a leash. There, she lays a single egg on it, which hatches quickly. The cockroach proceeds to stand placidly for the next week or so, unable to run away or defend itself, as it is slowly consumed alive by the wasp's offspring, which initially feeds on the outside of the unfortunate roach but eventually burrows into it and consumes its vital organs.

5

Shifting Baselines

An interesting aspect of these declines is that most of us have not noticed. The evidence suggests that insects, and also mammals, birds, fish, reptiles, and amphibians, are all now much less abundant than they were a few decades ago, but because the change is slow it is difficult to perceive. Among scientists it is now recognised that we all suffer from 'shifting-baseline syndrome', the phenomenon whereby we accept the world we grow up in as normal, although it might be quite different from the world our parents grew up in. Evidence is emerging that we humans are also poor at detecting gradual change that takes place *within* our lifetime.

Scientists at Imperial College, London, demonstrated these two related but different phenomena by interviewing villagers in rural Yorkshire. They asked them to name which birds were most common today, and which had been most common twenty years earlier, and compared their responses to the very accurate data we have on the actual birds that were common at these times. Not surprisingly, older people were better at describing which birds were most common twenty years ago. The scientists describe this as 'generational amnesia': for obvious reasons, young people simply don't know what the world was like before they were old enough to perceive it. More interestingly, although the older people could remember something about the birds of twenty years earlier, their description of which birds were most common back then was shifted towards what is found today. Their memory was imperfect, presenting a hybrid of accurate memory fused with recent observation, something the scientists describe as 'personal amnesia'. Our memory tricks us, playing down the magnitude of changes we have observed.

Of course, many people notice the common birds they see in their environment, but far fewer people pay any attention at all to insects. The only aspect of insect declines that has impinged on the consciousness of significant numbers of people has become known as the 'windshield phenomenon'. Anecdotally, almost everybody over the age of about fifty can remember a time when any long-distance daytime drive in summer resulted in a car windscreen so splattered with dead insects that it was necessary to stop occasionally to scrub them off. Similarly, driving country lanes at night in high summer would reveal a blizzard of moths in the headlights that has been likened to a snowstorm. Today, drivers in Western Europe and North America are freed from the chore of washing their windscreen. It seems unlikely that this can be entirely explained by the improved aerodynamics of modern vehicles.

I have an old recipe book for home-made wine, in which one recipe starts, 'First, collect 2 gallons of cowslip flowers ...' There must have been a time when this was a perfectly reasonable task to perform, but not within my lifetime. For me, cowslips have always been scarce, to the extent that spotting a few peeking from a hedge bank is a notable treat. The recipe provides a clue that our landscape was once much more abundant in flowers than it is today, but no one alive can remember this.

Although I never knew a time when there were vast numbers of cowslips, I think I can remember a time in the 1970s when there were many more butterflies. I am certain that flocks of lapwings were an everyday sight on farmland when I was a child, and that the unmistakeable springtime call of cuckoos could be heard anywhere I went in the countryside. Children of the new millennium are growing up in a world where butterflies, lapwings and cuckoos are scarce. They are never asked by their dad to scrub the windscreen of the car free of splatted insect remains after a summer drive. They almost certainly never spent a lunchtime in an overgrown corner of their primary school playing field, catching grasshoppers in their hands, because there usually aren't any. Just as I don't miss the fields of cowslips I never saw, so our children will never miss these things, as they never knew them. 'Normal' is different for every generation.

It seems probable that our children's children will grow up in a world with even fewer insects, and birds and flowers, than we have today, and they will think that normal too. They may read in books, or more likely online, that hedgehogs were once common, everyday creatures, but they will probably never experience the joy of hearing one snuffling about for slugs in a hedge bottom. They won't miss the flash of a peacock butterfly's wing any more than the present-day citizens of the USA miss the passing flocks of passenger pigeons, so vast that they once darkened the sky. They may be taught at school that the world once had great tropical coral reefs, teeming with fantastic and beautiful life, but these reefs will be long gone, no more real to them than mammoths or dinosaurs.

In the last fifty years, we have reduced the abundance of wildlife on Earth dramatically. Many species that were once common are now scarce. We can't be sure, but if one looks at the various studies from Europe, over various time periods and focused on different insect groups, it seems likely that we have lost at least 50 per cent or more of our insects since 1970. It could easily be as high as 90 per cent. Declines over the last hundred years are very likely to be much greater. North America is probably similar to Europe, since its farming methods are broadly similar, but we have far less certainty as to what has happened elsewhere in the world; it might be a bit better, or worse.

That we have such little certainty over rates of insect decline is scary, because we know that insects are vitally important as food, pollinators and recyclers, among other things. Perhaps more frightening, most of us have not noticed that anything has changed. Even those of us who can remember the 1970s, and who are interested in nature, can't accurately recall how many butterflies or bumblebees there were when we were children. Human memory is imprecise, biased and fickle and, as was found with the Yorkshire villagers, we are prone to revising our memories. You may have a vague, nagging feeling that there used to be more than just one or two butterflies on your buddleia bush, but you can't be sure. Perhaps it is just that a really good day stuck in your mind.

Does it matter, if we forget what once was, and future generations do not know what they have missed? Perhaps it is good that our baseline shifts, that we become accustomed to the new norm, as otherwise our hearts might break from missing what we have lost. A fascinating study of photographs of trophy fishermen returning to Key West, Florida with their catches from 1950 to 2007 estimated that the average size of the fish caught fell from 19.9kg to 2.3kg, but the smiles on the fishermen's faces are not any smaller. The fishermen of today would presumably be sad if they knew what they were missing, but they do not; ignorance is indeed bliss.

On the other hand, one could argue that we should fight to remember, and hold on to that sense of loss as best we can. Wildlife monitoring schemes can help us, by measuring the change. If we allow ourselves to forget, we will doom future generations to living in a dreary, impoverished world, not knowing the joy and wonder that birdsong, butterflies and buzzing bees can bring to our lives.

Leafcutter Ants

The leafcutter ants of South America form the largest and most complex societies on Earth, after humans.

As many as eight million individuals may live in a single colony, all sisters, caring for their mother, the queen, with the whole colony functioning effectively as what is sometimes called a 'superorganism'. Each worker has a specific role, and they have bodies to suit; for example, small workers tend to the brood in the nest, medium-sized workers forage for leaves, and large workers with greatly enlarged heads and fearsome jaws protect the nest from anteaters and other large predators. Some tiny workers hitchhike on the foragers, and defend them against parasitic flies, which seek to lay their eggs in crevices in the forager ants' heads. Somehow this entire complex enterprise functions without any individual being in charge. Like most animals, ants cannot themselves digest cellulose, the main constituent of plant material. Yet the foragers collect thousands of leaves per day, carrying them back to the giant underground nest along trails that meander across the rainforest floor. The leaves are carried back to underground chambers that hold a fungal garden, a culture of fungus that is tended meticulously by the workers, and fed upon chewed-up leaf pulp. The fungus can digest cellulose, and in return for being fed it produces small bundles of nutritious bodies known as staphylae, which are the main food of the ants. The fungi found in ant nests are found nowhere else; they cannot survive without the ants, and the ants would soon starve without them.

Part III

Causes of Insect Declines

What might be driving the global disappearance of insects? These days it sometimes seems as if there are almost as many theories as there are insects. Some are well supported by evidence, others less so but still plausible, and some are downright silly. Causes of the decline of wild bees have been discussed more than those of other insects, and although there is still debate, most scientists believe it results from a combination of man-made stresses, including habitat loss, chronic exposure to complex mixtures of pesticides, the spread of non-native insect diseases with commercial bee nests, the beginnings of the impact of climate change, and probably other factors too. There may well be drivers that nobody has yet recognised. Other insects probably face a similar range of difficulties. The causes of their decline are likely to vary from place to place; in short, this is complicated. However, if we are to halt and perhaps even reverse these declines, then we need to properly understand what is driving them, so we can work out what steps we need to take to make the world a more hospitable place for our insect brethren.

Amazon being simply knocked over and the timber burned *in situ*, smoke roiling across the blackened, skeletal branches, the most diverse ecosystem on Earth reduced to ashes. It was *déjà vu* to see the 2019 news footage of thousands of hectares of Amazonia once more on fire. Back in the 1980s, much of the tropical forest clearance was to create pasture for cattle destined to become burgers in cheap fast-food joints. There were protests and campaigns, but to little or no effect; the deforestation has continued unchecked over the decades. More recently, much of what is being cleared is replaced by monocultures of soya bean or palm oil, though some is still for cattle ranching.

Far from being curtailed, tropical deforestation today is happening faster than ever before, having increased in pace by somewhere between 10 and 25 per cent since the 1990s (estimating forest loss is not an exact science, but estimates become more accurate every year as satellite imagery improves). Tropical forests are currently being cleared at a rate of 75,000km² per year, or about 200km² per day, with considerably larger areas being damaged and degraded. The result is that an estimated 135 rainforest species go extinct every day, the large majority of them insects (though it must be noted that such estimates are necessarily very approximate). On Easter Island, deforestation continued until every last tree was gone, and the soils washed away into the sea. Just as the Once-ler cut down all of the truffula trees despite the warnings of the Lorax, we continue to deforest our planet, knowing it is an idiotic thing to do. We are committing ecocide on a biblical scale. I am not in any way religious, but if you are, consider this; do you really think God created wonderful diversity and gave us dominion over it so that we could exterminate it? Do you think He or She is pleased with what we have done?

Of course, it is not just the tropical forests. Temperate and boreal forests are being destroyed too, so that globally we are suffering a net loss of about one billion trees per year. Between 2000 and 2012 we lost 2.3 million square kilometres of forest worldwide – an area greater than the combined size of the UK, France, Germany, Spain, Portugal, Belgium, the Netherlands, Italy, Switzerland, Austria, Poland, Ireland and the Czech Republic. If it was all gathered together, one could walk from John O'Groats, at the north-eastern tip of Scotland, south to Gibraltar and then east to Warsaw, without ever passing beneath the shade of a tree. Only 6.2 million square kilometres remain of the 16 million square kilometres of forest that once clothed the Earth.

Other wildlife-rich habitats are also being damaged or destroyed: lakes and rivers have become polluted and degraded, marshlands drained, peatlands dug up or drained, and valleys flooded by dams for hydroelectric schemes or irrigation. In China entire mountains are being bulldozed, their peaks demolished and used to infill valleys, creating flat plains of land needed for the expansion of cities. In Tokyo, shallow coastal areas are being infilled with garbage, creating brand new islands out of trash that are being used for buildings and golf courses.

At a global scale, the ongoing loss of pristine habitats and their replacement with massively simplified, anthropogenic ones is at present probably the single biggest driver of wildlife declines, including declines of insects (although habitat loss may soon be eclipsed by the ravages of climate change). Within Western Europe, habitat loss has taken a different form. Here, almost all the pristine, natural habitats were lost centuries ago. In the UK, even our few remaining ancient woodlands have been managed in one way or another for millennia – they are not truly wild, wonderful though they may be. Yet despite 8,000 years of human occupation (indeed, perhaps *because* of this occupation), until about 1900 we still had large areas of habitats that were rich in biodiversity. Aside from mature woodlands there were the chalk downs, awash with flowers and butterflies, lowland hay meadows in which corncrakes nested, coppiced woodlands where heath fritillary and white admiral butterflies soared, heathlands where sand lizards hunted for grasshoppers. These were all man-made habitats, created by traditional land management practices: hay cutting, grazing and coppicing. These are relatively gentle or infrequent ways of managing the land, practices that wildlife can accommodate to, and even benefit from. Trees are coppiced (chopped down to ground level, encouraging them to sprout back) only once every ten or twenty years, so that coppiced woodlands are a matrix of open glades and different-aged stands of trees, supporting a diversity of life. Occasional grazing of the downs and annual hay-cuts of meadows kept saplings from establishing and coarser grasses under control, allowing flowers to flourish so long as the numbers of grazers were kept low. Over millennia, these habitats became key to the survival of much of the wildlife of Europe.

Today, it is the loss of these man-made habitats that is a major driver of wildlife declines in Europe, primarily driven by the rapid change in how we farm. Historically, less intensive farming practices resulted in a patchwork of habitats that were favourable to bees and other insects; aside from the meadows and chalk downlands there were fallow fields rich in flowering weeds, and flowering hedgerows separating the small fields. In the 1920s, the UK had about 3 million hectares of downlands and hay meadows, but more than 97 per cent was lost during the twentieth century. Most of it was replaced by arable crops or silage fields, habitats which typically support close to zero biodiversity.

The shrill carder bumblebee is now the UK's rarest bumblebee, but a hundred years ago it was a familiar sight and sound in the south of Britain. A colourful little bumblebee with yellow and greyish stripes and a red bottom, the species gains its name from its unusually high-pitched buzz, which is often the best giveaway that there is one foraging nearby. The shrill carder bumblebee is a species of flower-rich meadows, where it loves to feed on flowers such as red clover, red bartsia, kidney vetch, knapweed and viper's bugloss. The loss of almost all our flower-rich grasslands during the twentieth century nearly drove this handsome bee extinct, and now it clings on in a handful of sites, including Pembrokeshire, the Somerset Levels and the

Thames Estuary. Twenty years ago I saw some on Salisbury Plain, but that population seems to have died out. Some of the best populations that remain are on brownfield sites near the Thames to the east of London, abandoned industrial areas strewn with burned-out cars and rubbish, which have become rich in flowers and now serve as one of the last refuges for the shrill carder.

Many other meadow creatures have undergone massive declines in the face of the loss of their meadow home: corncrakes, scabious mining bees, chalkhill blue butterflies, great yellow bumblebees, wart-biter cricket and greater butterfly orchids, to name but a few. The full list would go on for pages.

Flowery meadows are not the only habitat we have lost in the UK. In the wake of the Second World War, government subsidies were introduced with the intention of increasing food production and farm efficiency. These included payments for hedgerow removal, resulting in the loss of 9,500km of hedges per year. It is estimated that we lost a little over half of all our hedgerows between 1950 and the millennium. Eighty per cent of our lowland heaths have been lost since 1800, along with 70 per cent of our farmland ponds, with most of the remaining ponds ruined by heavy pollution, particularly by fertiliser run-off.

Modern farming, in the UK and elsewhere in the developed world, has been shaped by global agribusinesses and government policies, and is typified by large farms with large fields, often managed by external contractors, maintained so far as possible as near-perfect monocultures by high inputs of pesticides and fertilisers. The drive has been to maximise crop yields, justified by arguments that we must ensure 'food security', born out of the food shortages experienced during the Second World War. Proponents of industrial farming argue that it is the only way to avoid starvation and famine. In going down this route we have created a landscape that produces much more food, more cheaply, than it used to, but it is one that provides employment for very few people and is largely inhospitable to wildlife. From the perspective of insects, modern intensive farmland typically offers few opportunities. Flowers are scarce, so for butterflies, moths, bees and hoverflies there is no nectar or pollen. Weeds that might have provided food for caterpillars, froghoppers or beetles are few and far between. The sparse hedgerows, often flailed low to the ground, offer little refuge for insects seeking a nest site or place to hibernate. Any insect that does find food or shelter will have to survive exposure to repeated insecticide sprays.

The low price of food on the supermarket shelves that we have become accustomed to does not reflect the true environmental costs of its production. For example, nitrates from fertilisers and metaldehyde from slug pellets used on arable fields wash into and contaminate our streams and rivers. Water companies extract water from rivers for human consumption, and have to spend large sums on trying to remove these pollutants – metaldehyde in particular is very hard to remove, so that, despite their best efforts, there is usually still some in our drinking water. We or our children will all pay, in the long run, for the erosion and degradation of farmed soils,* for the greenhouse gases released by farming activities (which account for about 25 per cent of all emissions), and for the loss of pollinators and other insects.

The combined effect of the loss of pristine habitats (mainly in developing countries) and the loss of semi-natural habitats (mainly in developed countries) is that wildlife, globally, finds

---

* Current estimates suggest that we are losing about 75–100 billion tons of topsoil each year from the surface of the globe, with particularly high rates in China and India, and the USA not too far behind. Even New Zealand, a country that many might think of as relatively environmentally aware, is losing an estimated 192 million tonnes of soil per year, with much of this coming from overgrazed pastures. With New Zealand's human population standing at just 4.8 million, this equates to 40 tonnes of soil lost per person per year. The global average is roughly 10 to 15 tons of soil lost per person on the planet – troublesome figures, given that soil takes thousands of years to regenerate. Poor farming practices leave soil exposed, with the organic matter oxidising to add to carbon dioxide emissions, and much soil washing or blowing into rivers and oceans, where it causes silting and pollution.

itself increasingly squeezed into small, fragmented and isolated patches of surviving habitat 'islands', be they pieces of rainforest that have so far escaped the chainsaw, or those nature reserves sampled by the Krefeld Society in Germany. It is commonly imagined that the wildlife lucky enough to find itself in a nature reserve is safe, but the German study shows unequivocally that this is not so. In the Krefeld study, the 76 per cent decline in insect biomass between 1989 and 2016 occurred on nature reserves that remained intact, in more or less the same state and carefully managed for wildlife, throughout the study. Although the German data do not provide clear evidence for the cause of the decline, we can make some educated guesses. Those German nature reserves, as with those elsewhere, tend to be hemmed in by inhospitable habitats. The study was of flying insects (the large majority of insects fly), and any flying insect is likely to fly out of the reserve before long. If it then finds itself in a landscape where it cannot survive, owing to a lack of food or high levels of pesticides or other factors, then unless it has the good sense to retrace its steps it is probably doomed. The surrounding landscape acts as what population biologists call a 'sink', into which organisms go and rarely return. Unless the island population can breed fast enough, this steady haemorrhaging of individuals into the wider landscape will likely lead to local extinction.

We know that some insects do have the common sense to stay put most of the time. Adonis and small blue butterflies, for example, both tend to stay close to where they are born throughout their adult life, a wise strategy in the modern world. Unfortunately, even then they are not safe, for small island populations become inbred over time, losing genetic variation and becoming less healthy and adaptable as a result. Unless regular migrants make their way across the surrounding wastelands, bringing with them fresh genes, sooner or later the population is doomed.

If all this weren't enough, small island populations are prone to go extinct just due to bad luck. Populations of insects vary a lot from year to year, particularly because of the vagaries of the weather. A bad storm, a flood or a summer drought might be all that's needed to wipe out a small population that may have clung on for decades. Once a particular species has been lost from a nature reserve it is unlikely to be able to recolonise, unless there happen to be healthy populations on nearby reserves which can act as a source of migrants. As habitats become more and more fragmented and isolated, recolonisation happens less and less frequently.

There is a final, insidious factor at play. Putting a fence around a nature reserve does not stop pesticides drifting in on the wind or seeping in through the groundwater. It does not prevent deposition of oxidised nitrogen compounds, the products of burning fossil fuels, which fertilise the soils and alter the plant community. And of course it does not halt the advances of climate change, which over time may render a particular place climatically unsuitable for some – or eventually all – of its current occupants.

One way or another, if we take large expanses of intact habitat and remove most of it, leaving only small scraps (as we have for woodlands, heathlands and chalk downlands, for example), we would expect the number of species living on those small scraps to fall over time as populations fizzle out, one by one. This may happen decades after the islands were first created, as we watch the gradual, inexorable payback of an extinction debt. As David Quammen, the American science writer, put it in his excellent book *Song of the Dodo:*

Let's start by imagining a fine Persian carpet and a hunting knife. The carpet is 12 feet by 18, say. That gives us 216 square feet of continuous woven material. Is the knife razor-sharp? If not, we hone it. We set about cutting the carpet into 36 equal pieces, total them up – and find that, lo, there's still nearly 216 square feet of recognisably carpet-like stuff. But what does it amount to? Have we got 36 nice Persian throw rugs? No. All we're left with is three dozen ragged fragments, each one worthless and commencing to come apart.

This is what is happening in Germany, and probably all over the world.

This phenomenon has direct relevance to a major scientific controversy, often called the 'sharing-sparing debate', in which 'sharers' advocate trying to integrate human activities such as growing food with supporting biodiversity (for example with small, organic, eco-friendly farms), while 'sparers' argue for using some land as intensively as possible (for example for industrial farming), so that the remainder can be set aside for nature. But what the German study shows is that setting aside land for nature does not seem to work, at least when the areas protected for nature are small and surrounded by areas used for industrial farming.

Overall, we can be quite certain that past habitat loss, of pristine habitats such as tropical rainforests as well as man-made habitats such as hay meadows and lowland heaths, is right up there as one of the biggest drivers of insect declines to date. A top priority has to be to find ways to curtail any further losses, and perhaps even begin restoring some habitats to their former glory.

Orchid Bees

In the steamy jungles of Central and South America live the orchid bees, a family of fabulous metallic-green, gold or blue bees that glint like jewels in the tropical sunshine as they zoom from flower to flower. Male orchid bees live up to their name by spending much of their time visiting orchids, but these orchids provide no nectar, and neither do the bees collect pollen. Instead, they use brushes on their forelegs to collect aromatic chemicals from the orchids, which they comb into and store within their greatly enlarged and hollow hind legs. They are perfume collectors. Males then gather at display sites, which females visit to find a mate, apparently choosing males on the basis of the quality and quantity of the orchid scents that they have collected.

Orchids have an unusual pollination system. Instead of producing scattered grains of pollen as most flowers do, each flower produces one or two pollinia: dense balls of pollen with a sticky stalk that attach to a visiting insect. The orchids visited by male orchid bees are entirely dependent on them for pollination, something first described by Charles Darwin, although he thought the bees were female. The structure of the orchid flower is carefully arranged so that, while the bees are busy brushing up floral scents, their head or thorax comes into contact with the stalk of the pollinia, and the whole structure becomes stuck to them, a pair of bright yellow pollen balls that the bee cannot dislodge. When the bee visits subsequent flowers to collect more scents for their perfume, some of the pollen is transferred, fertilising the flower. If all goes well, this unique symbiosis ensures that both orchid and bee get to procreate.

7

The Poisoned Land

Since the dawn of agriculture 10,000 years ago, our crops have been attacked by diseases, or by animal pests ranging from aphids and locusts to pigeons and elephants. As the human population grew, and the extent of arable farmland grew to match, these pest problems worsened, for the more of a crop you grow the more likely it is that a pest will find it. So far as we know, for the first 5,000 years or so of agriculture farmers relied upon prayer and ritual sacrifice as their primary means of crop protection. In ancient Egypt, for example, slaves were sacrificed to appease Renenutet, goddess and protector of the pharaoh's harvest, while the Aztecs sacrificed children to their rain god, Tlaloc. My guess is that these bloody ceremonies were probably not successful. Nonetheless, it remains common for farmers and rural communities to pray for divine aid, although more pragmatic means of pest management have also long been practised. As far back as 4,500 years ago, for instance, farmers are thought to have applied sulphur to their crops to kill insect pests. The Chinese were using mercury and arsenic compounds for controlling body lice 3,200 years ago, and probably also sprinkled them on their crops. Plant extracts, for example from the 'Pyrethrum daisy' *Chrysanthemum cinerariaefolium*, have been applied as insecticides for at least 2,000 years. Using chemical pesticides is certainly not new.

However, up until the 1940s the pesticides we used were either naturally occurring organic compounds such as pyrethrum or nicotine, usually extracted from plants, or inorganic compounds such as copper sulphate, mercury salts, cyanide, arsenic or sulphuric acid. Some argue that naturally occurring compounds are more benign than modern synthetic alternatives, but this is clearly nonsense. There is nothing environmentally friendly about mercury or arsenic. We do not have figures for how much of these chemicals were used, but it is a reasonable guess that the overall amounts were very small, for most farmers simply couldn't afford them or had no access to them.

All this was to change with the advent of industrial chemistry. The industrial-scale manufacture of chemicals began in the eighteenth century, with the manufacture of sulphuric acid, bleach, and later of soda used in the production of glass and textiles. During the nineteenth century there was a massive expansion of the chemical industry, producing dyes, vulcanised rubber, fertilisers, soaps and the first plastics. Yet it wasn't until the twentieth century that the new industry turned its attention to developing novel, synthetic pesticides.

DDT (dichlorodiphenyltrichloroethane) was the first man-made compound found to have insecticidal properties. The discovery of a Swiss chemist, Paul Hermann Müller, in 1939, DDT attacks insects' nervous systems, causing nerve signals to fire off repeatedly, leading to twitching, tremors, seizures and ultimately death. Widely used during the Second World War to control malarial mosquitoes that were plaguing the Allied troops fighting in Asia, DDT had also, by the end of the war, become widely and cheaply available for domestic and farming use. In 1947 one manufacturer ran an advert in *Time* magazine that showed smiling cartoon farm animals and a rosy-cheeked housewife all singing 'DDT is good for me-e-e!', along with the claim that 'DDT is a benefactor of all humanity'. A short film from the same year shows a British colonialist dousing a bowl of porridge in DDT and then eating it, in a bid to persuade the local people in East Africa that the new chemical was harmless to them (the audience seemed unimpressed). Meanwhile, Paul Hermann Müller was awarded the Nobel Prize in 1948 for his discovery.

In parallel, a German scientist, Gerhard Schrader, synthesised another chemical, parathion, during the 1940s. This too was highly toxic to insects, attacking the nervous system, preventing breakdown of neurotransmitters and leading to disorientation, paralysis and death. The company Schrader worked for, I. G. Farben, also developed and manufactured Zyklon B for use in the gas chambers, and it is likely that his work was part of research intended to develop nerve agents for use against people.

By chemical tinkering, various related compounds were soon developed. DDT and its relatives are known as organochlorides, and include aldrin and dieldrin, while parathion spawned dozens of organophosphate compounds including malathion, chlorpyrifos and phosmet. These new chemicals were cheap and highly effective at killing insect pests, giving bumper crops – at least to start with – and so they were enthusiastically adopted by farmers. A global industry sprang up in developing, manufacturing and distributing an ever-growing range of these toxins. The 1970s and 1980s saw the introductions of many new families of pesticides, including avermectins (fed to livestock to kill parasites), *Bacillus thuringiensis* sprays (insecticidal toxins extracted from a type of bacteria), and the triazoles, imidazoles, pyrimidines and dicarboxamide fungicides. In the 1990s yet more new products came to the market, including the brand-new neonicotinoid family of insecticides, and also spinosad and fipronil. Today, about 900 different 'active ingredients' – types of chemical that are toxic to some sort of pest – are licensed for use in the USA, and about 500 in the EU. Over the last eighty years or more farming has become steadily more addicted to chemical inputs, a process that continues to the present. According to official government statistics, UK farmers treated 45 million hectares of arable land with pesticides in 1990. By 2016, this had risen to 73 million hectares. The actual area of crops remained exactly the same, at 4.5

million hectares. Thus each field was, on average, treated with pesticides ten times in 1990, rising to 16.4 times in 2016, a rise of nearly 70 per cent in just twenty-six years.

It was in 1962, only eighteen years or so after DDT entered farming use in the post-war period, that Rachel Carson published her groundbreaking book *Silent Spring*, which put a spotlight on the early generations of pesticides and the growing evidence that they were not as benign as had been naively

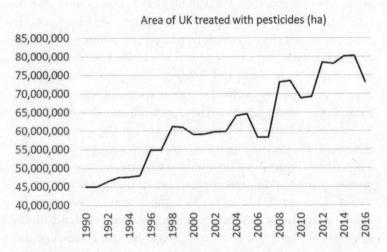

The area of farmland treated with pesticides each year in the UK: Every year, farmers make more pesticide applications to their crops. The chart shows the official government figures [from https://secure. fera.defra.gov.uk/pusstats/] for the total area of crops treated with pesticides each year in the UK (74 million hectares in 2016). This area increased by 70 per cent between 1990 and 2016. Given that there are only about 4.5 million hectares of arable and horticultural land in the UK, and that this area remained almost unchanged over this period, these figures mean that each field or orchard in the UK is now, on average, treated about sixteen times annually. It should be noted that this could be the same pesticide applied sixteen times, or sixteen different pesticides each applied once, or some combination of the two. These data do not include pesticides used by farmers for veterinary purposes, such as the avermectins routinely given to livestock to protect them against parasites.

assumed. The problem was that pest insects had rapidly evolved resistance to the new pesticides, so farmers were having to apply them in ever-larger quantities, and the early bumper crops could not be sustained. Natural enemies of pests, however, such as predatory wasps and beetles, tend to breed more slowly than their prey, so are slow to evolve resistance, and were hit harder by the pesticides. Without those natural enemies, pest problems became steadily worse, and new pests emerged: insects that had previously been kept in check by their predators. It also became apparent that DDT and its relatives persist for decades in the environment, accumulating up the food chain – a caterpillar being eaten by a songbird which will in turn be eaten by a falcon, and so on – so that both predators and we humans end up with large quantities of them lodged in our fat deposits. Higher doses lead to death, with lower doses causing cancers, spontaneous abortions and sterility. Raptors such as peregrine falcons and bald eagles were particularly affected, with exposure leading to eggshell thinning, so that most of their eggs were accidentally broken before they could hatch.

Following the publication of *Silent Spring*, Rachel Carson came under personal attack from the agrochemical industry and its lobbyists, labelled, among other things, as a fanatic and a communist. Industry launched a counter-offensive, publishing leaflets, lodging complaints and threatening legal action against the publishers of *Silent Spring*. Ultimately, Carson won her battle, although sadly she did not live to see it, dying of cancer in 1964. DDT was banned in 1972 in the USA, in 1978 in Europe, and globally in 2004, aside from some limited use in malarial control. Yet screening of soils and rivers in Europe continues to find residues of the chemical. Without in any way wanting to undermine the undoubtedly huge benefits of feeding human babies on breast milk, it is concerning that human milk is still very often contaminated with DDT and its relatives, typically containing ten to twenty times more organochloride insecticides (and also far more polychlorinated biphenyls, PCBs) than cow's milk (based on studies from diverse locations including

Australia, Mexico, Ukraine and the Canary Isles). By feeding on humans, human babies are of course at the very pinnacle of the food chain. DDT is not quite so good for me-e-e after all.

So much for DDT, but the organophosphates invented by Gerhard Schrader also turned out to be extremely hazardous to farmers' health, as might be expected for chemicals emerging from nerve agent research. In particular, farmers who dipped their sheep in these chemicals suffered a broad range of acute and long-term health problems. In developed countries most – but not all – organophosphates have now also been banned, although they are very widely used to this day in developing countries.

These days, the proponents of pesticide use often argue that modern pesticides are much safer for people and the environment than the older, banned pesticides – a view which seems to have prevailed, unquestioned, for decades. There is a terrible irony to the fact that conservationists and independent scientists appear to have accepted that the problem had been solved. Rachel Carson, they believed, had won. A search of global databases of scientific publications, using the key words 'wildlife' and 'pesticide', reveals that just twenty-nine papers were published between publication of *Silent Spring* in 1963 and 1990 (in contrast, 1,144 papers have been published on this topic post 1990). Essentially, both conservationists and scientists took their eye off this particular ball. Carson may have won a battle, but not the war.

The re-emergence of concern over the impact of pesticides on the environment can be traced back to the 1990s, when French beekeepers began to complain that their honeybee colonies were dying in the vicinity of sunflower crops that had been treated with a new insecticide, imidacloprid. Imidacloprid was the first of a new family of chemicals, the neonicotinoids – a name that meant nothing to most of us at the time, but was to become notorious for its links to bee declines. Like DDT and the organophosphates, neonicotinoids are neurotoxins, attacking the brain of an insect, but they are far more potent than their predecessors: the dose of DDT needed to kill a honeybee is 7,000 times higher than the dose of imidacloprid.

French farmers are notorious for their militant behaviour – only a few years earlier French farmers had set light to a lorry load of British sheep in protests over cheap imports – yet for a long time the French beekeepers were largely ignored, despite staging a march in Paris dressed in full beekeeping gear. A decade later, however, honeybees in North America started to die in large numbers in a phenomenon dubbed 'colony collapse disorder'. In many cases the adult bees were vanishing, abandoning their brood to die, in a sort of adult bee variation of 'the Rapture'. The statistics were huge and shocking: 800,000 honeybee colonies, representing nearly one-third of all the honeybee colonies in North America, were lost in the winter of 2006/7, and nearly as many the next year. There was a storm of media interest and wild speculation that bees might be about to go extinct. The exact causes were not clear, with diverse theories of varying credibility blaming viral diseases, a parasitic blood-sucking mite named *Varroa*, mobile phone signals, alien abduction, chemtrails, poor diet, and pesticides. Soon, however, research programmes sprang up in laboratories across North America and Europe to try and identify the causes of the 'bee crisis'.

And it wasn't just France and America, for in the spring of 2008 thousands of honeybee hives died in Germany from mass poisoning. These deaths coincided exactly with the timing of farmers sowing maize seeds, almost all of which had been coated in a neonicotinoid insecticide. Subsequent investigation revealed that the seed-coating process had been faulty, causing the insecticidal coating to come loose to create a toxic dust cloud when the seeds were being sown into the ground. Although the fault was swiftly remedied, the attention of those scientists such as myself who were trying to understand the causes of bee declines was finally drawn to neonicotinoids (we scientists can be slow on the uptake). By that time almost all of the arable crops in Europe and North America were being treated with neonicotinoid seed dressings. Could it be that those French beekeepers had been right all along?

Neonicotinoids are systemic, meaning that they travel to all parts of the plant. When they're used as a seed dressing, the intention is that the seed coating dissolves in the damp soil once the seed is sown (the chemicals are water-soluble), and, as the seed germinates and grows, that it sucks up the toxin, which then spreads throughout the plant and protects it against insect pests. This sounds pretty clever – the farmers buy the seed pre-coated, and then have to do nothing more to protect their crops. What should have been obvious, but does not seem to have worried anybody when these new chemicals were being introduced, is that anything that spreads to all parts of the plant will spread into the pollen and nectar too. And of course crops such as oilseed rape and sunflowers require pollination and are popular with many types of bees, all of which might be dosing themselves with insecticide when the crops bloom.

In the early 2000s, analytical tests revealed that the nectar and pollen of treated crops did indeed contain traces of neo-nicotinoids, albeit at low levels in the region of a few parts per billion. A debate began as to whether such concentrations were sufficient to do any harm. The manufacturers of neoni-cotinoids, the agrochemical giants Bayer (an offshoot of the aforementioned I. G. Farben) and the Swiss company Syngenta, strenuously denied any link between their pesticides and bee declines. Experiments were needed to test whether bee colonies were harmed by these concentrations, but it takes years to raise funds, conduct experiments, analyse them, and publish them in scientific journals.

My own research group, then based at the University of Stirling in Scotland, set about studying whether bumblebee colonies feeding on oilseed rape crops treated with the neo-nicotinoid imidacloprid might suffer any harm. This might sound easy, but in reality it is not a trivial undertaking. The ideal field experiment would have involved having many fields of oilseed rape, each randomly allocated to being treated with the pesticide or left untreated as 'controls'. Bumblebee colonies would then be placed next to them, and their health measured over time. Many fields are needed, as *replication* is vital in ecological studies, as there is always unexplained variation: every field, and every bee colony, is slightly different. Having lots of replicates enables us to discern patterns due to experimental manipulations – in this case, exposure to the pesticide – from this background noise. Each field would have to be at least 2km from the other fields so that bees would not fly between them, for if they did then bees next to control fields might still be exposed to the pesticide. Ideally, the entire landscape would be free of any other crops treated with neonicotinoids.

To set up such an experiment would require large amounts of money, but we had no sponsor, and no funds, so of necessity we took a different approach. Instead of placing bumblebee colonies next to treated or untreated fields (ideally with many replicates of each), we spiked the food of bumblebee colonies with imidacloprid, to mimic concentrations found in the nectar and pollen of a seed-treated oilseed rape crop. These concentrations were extremely low – just 6 and 0.7 parts per billion of imidacloprid in the pollen and nectar respectively. The bees received this contaminated food for two weeks, to simulate what might happen if a bee nest were near a treated oilseed rape field, and thereafter the seventy-five nests were placed out on the university campus and left to look after themselves. We monitored their weight every week, and counted how many new queens the nests produced. Bumblebee nests die at the end of the summer, but if all has gone well they leave behind new young queens which will start their own colonies the following spring.

When we collated and analysed the data, the results were quite clear. Nests exposed to the pesticide were quite a bit smaller than control nests (meaning those nests given uncontaminated food), and produced 85 per cent fewer new queens. That, of course, would mean far fewer bumblebee nests the following year. The results were concerning, and we were

delighted when the high-profile journal *Science* agreed to publish the study in early 2012.

The work was published alongside a study from a French research group led by Mickaël Henry in Avignon, who had found
that neonicotinoids impaired the navigational ability of honeybees,
so that they often got lost on their way back to their hive after a
foraging trip. It seemed to us that the two studies provided clear
evidence that these neonicotinoid seed dressings could indeed
harm bees, and the results were reported around the world.

At the time I was rather naïve. New to the controversial
world of pesticides, I was unprepared for the backlash from
pesticide companies, who set out to undermine our findings.
Our experiment was not realistic, they argued, as we spiked
the food rather than allowing bees to naturally feed on the
crop, effectively forcing the bees to feed on the contaminated
food. It was said that both our study and the French one used
unrealistically high doses of the pesticide, even though the doses
we used were taken straight from published analyses of the
concentrations of neonicotinoids in the pollen and nectar of
oilseed rape crops. Defamatory online articles appeared which
attempted to discredit me personally as a scientist, claiming that
I was a 'scientist for hire', willing to take money and fabricate
evidence to support any agenda (though in fact our research
had not been funded by anyone).

Fortunately the European Parliament took our results sufficiently seriously to ask the scientists at the European Food
Safety Authority (EFSA) to look into the subject and report
back. It took them most of a year to review all of the evidence,
but in 2013 they advised that the use of neonicotinoids on
flowering crops – those such as oilseed rape and sunflowers
that attract pollinating insects, as opposed to wind-pollinated
crops such as wheat and barley – did indeed pose a significant
risk to pollinators. To its great credit, the European Parliament
acted quickly, proposing a ban on the use of neonicotinoids on
flowering crops, and although to my embarrassment the UK
initially opposed this, the ban became law in December 2013.

In the meantime, the science was moving on. New experiments were published, some of them large-scale field trials of the sort that we had not been able to afford. A huge Swedish study led by the ecologist Maj Rundlöf and published in 2015 found that bumblebee nests placed next to actual treated fields of oilseed rape performed much more poorly than when placed next to untreated fields. It found that queen reproduction was impaired by almost exactly the 85 per cent we had reported three years earlier. At the same time as studying bumblebees, these researchers also looked at the effects on solitary mason bees and honeybee colonies. The mason bees entirely failed to breed next to the treated fields, but there was no significant effect on the honeybee colonies.

The Swedish study was followed two years later by an even larger international study, with replicated fields in the UK, Germany and Hungary. This huge project was carried out by Ben Woodcock and others from the UK's Centre for Ecology & Hydrology (CEH). It was funded to the tune of £2.8 million by the agrochemical industry itself, with the protocols agreed in advance by them, a last-ditch attempt to overturn the European ban on the use of neonicotinoids on flowering crops. This study too found that neonicotinoids on oilseed rape were harmful to bumblebee colonies and solitary mason bees. In the UK and Hungary the honeybees were also measurably harmed by the pesticides, but not in Germany, where the bees seemed to feed mainly on wildflowers away from the crop.* The funders, Bayer and Syngenta, quickly distanced themselves from the work, criticising the very methods that they had agreed to in

---

* A common thread emerging from these and other field studies is that wild bees such as bumblebees and mason bees seem to be more badly affected by these pesticides than honeybees. We cannot be certain why, but a popular theory is that the much larger size of honeybee colonies, which may have 50,000 workers, means that they can cope with losing some to pesticides. In contrast, bumblebee colonies have at most a few hundred workers, while in mason bees the female single-handedly looks after her nest, so if anything happens to her then the nest is finished.

advance, claiming the results were inconclusive, and accusing the project team of cherry-picking and misrepresenting data. Ben Woodcock struck back, saying in a media interview, 'I don't really appreciate having them accuse me of being a liar.'

To this day, the manufacturers of neonicotinoids maintain that their insecticides are effective at killing insect pests but are entirely benign to bees and other beneficial insects, despite the mountain of evidence to the contrary. Their stance reminds me of nothing so much as 'Doublethink': the ability simultaneously to accept two contradictory beliefs, which in George Orwell's *Nineteen Eighty-Four* is a capacity expected of all loyal Party Members.

While Ben Woodcock and Maj Rundlöf were running their huge field trials, other scientists were looking in more detail at the effects of neonicotinoids on the behaviour and health of individual bees (rather than colonies). It only takes a very small amount of a neonicotinoid to kill a bee outright. Toxicity is often measured as an LD50, which stands for 'lethal dose 50 per cent': the dose that kills half of the animals that receive it. The LD50 for most neonicotinoids is about four-billionths of a gram per honeybee, which is not much by any standards. However, evidence began to emerge that even smaller amounts could have subtle but important 'sublethal' effects. We already knew from Mickaël Henry's 2012 study into the effects of imidacloprid on bee navigation that giving honeybees one-third of the LD50 dose could reduce their ability to find their way back to their hive. This makes intuitive sense: any of us given a sublethal dose of a neurotoxin might well feel dazed and confused and take a wrong turning on the way home. For a honeybee, whose job it is to spend all day gathering food by flying backwards and forwards between her hive and patches of flowers, getting lost is disastrous. She is no longer contributing to the hive, and she will not survive long outside. Could this be the explanation for the 'bee Rapture'?

To make matters worse, new studies emerged suggesting that tiny sublethal doses of neonicotinoids also have other harmful effects. For example, just one part per billion of neonicotinoids in food impairs the immune system of bees, leaving them susceptible to infection by unpleasant diseases such as deformed wing virus (the symptoms of which include shrivelled wings, making the bees unable to fly). Tiny doses of neonicotinoid, whether received as a developing larva or as an adult, seem to impair the ability of bees to learn and remember which flowers are most rewarding, a vitally important skill for colony success. These sublethal doses also reduce egg laying and life expectancy of queen bees, reduce fertility of male bees, and reduce the amount of time adults spend looking after the brood. None of the regulatory tests normally carried out to check whether a new pesticide is likely to be harmful to bees include looking at these sublethal effects.

You might think that the EU ban on use of neonicotinoids on flowering crops would have been the end of this particular matter, at least in Europe: if bees aren't exposed to these chemicals then neither lethal nor sublethal effects are of any consequence in the real world. Unfortunately, it turns out that it is not so simple. Gluing a pesticide onto the crop seeds seems like an efficient way of using them (so long as they are stuck on properly). Previously, most pesticides were sprayed from a boom mounted on a tractor, with the potential for the spray to drift beyond the crop, into hedgerows, gardens, or nature reserves, for example. Using them as a seed dressing was promoted as providing much better targeting of the pest, and on the face of it this seems pretty convincing and was widely accepted as fact. Sadly, it turned out to be untrue.

In 2012, a study was published by the US scientist Christian Krupke which found neonicotinoids in wild dandelions growing on farmland near treated crops. When I read this paper it set off alarm bells. If the pesticide was glued to the crop, how was it getting into wildflowers nearby? Digging around in the more obscure scientific literature, I found a study published by Bayer's own employees, who had quantified what proportion of the seed dressing was taken up by the crop. It varied greatly between crop types, from about 1 to 20 percent of the active ingredient, with an average of just 5 per cent. For comparison, spraying pesticides from a tractor can easily get 30 per cent or more of the active ingredient onto the crop. So if an average of 95 per cent of the neonicotinoid seed coating was not going where it was supposed to, where on earth was it going?

Research from Italy then revealed that, even when the insecticide was properly glued to the seed, about 1 per cent of the pesticide was blown off as dust during the seed-drilling process. The Italian researchers found that this dust was deadly to any honeybee flying nearby.

This still left 94 per cent of the chemical unaccounted for. Of course, it was fairly obvious where it was going – into the soil and the groundwater. If the bulk of the neonicotinoid was not being sucked up by the crop, and was not blowing around in the air, presumably it was left behind in the soil. This then raised further questions: was it doing any harm in the soil, for example to all the myriad little creatures that help to keep soil healthy? How long was it lasting in soil? Was it also contaminating the soil water and seeping into waterways?

Then, in 2013, I stumbled across evidence tucked away in a massive and obscure EU report on imidacloprid, which showed that this chemical broke down very slowly in soil, so that it accumulated over time if a seed-treated wheat crop was sown every year. This was a six-year study conducted by Bayer, beginning back in 1992, but its significance seemed to have been overlooked by the pesticide regulators for more than fifteen years. I only discovered it myself when I received an anonymous email from somebody in the United States, directing me to what turned out to be a 700-page document.

Persistence of a pesticide is highly undesirable, for if it takes years to break down in the environment then it is much more likely to come into contact with something it isn't meant to harm. This becomes much worse if the chemical is so persistent that it accumulates with each successive application, so that the amount of toxin in the environment gets higher and higher over

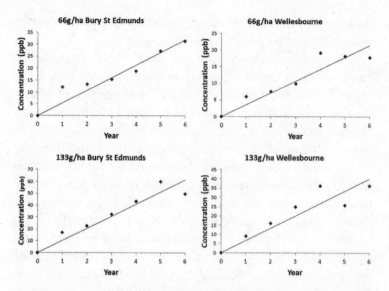

Accumulation of a neonicotinoid insecticide in soil: Levels of the neonicotinoid imidacloprid detected in soil into which treated winter wheat seeds were sown each autumn (1991–6). The two study sites were both in England. Treatment rates were 66g or 133g of imidacloprid/ha except in the first year, which was 56g in Bury St Edmunds, and 112g in Wellesbourne. The data are from the EU Draft Assessment Report for Imidacloprid, 2006, and show beyond any doubt that levels of the chemical build up over time. Yet, somehow, the report concludes from these data that 'the compound has no potential for accumulation in soil'.

time. The long persistence of DDT was one of the main reasons why it was eventually banned. If neonicotinoids were building up in soils, then any soil organisms would be heavily exposed to them throughout the year. Given that neonicotinoids are water-soluble, one might expect them to leach from crop fields downhill into neighbouring land, and into ditches, streams and rivers. This might also explain how the pesticides had come to be in dandelions in Christian Krupke's study – if the soil was contaminated, presumably the roots of wildflowers could take them up just as readily as crops? If they were in wildflowers, then the EU ban on their use on flowering crops might not be enough to prevent bees from being exposed.

When it came to my own work, shortly after we published our 2012 paper on the effects of neonicotinoids on bumblebee colonies I moved from Scotland to the University of Sussex on England's southern coast. The move was followed by an unusual run of good luck with grant applications. More or less simultaneously I got funding from DEFRA and from the UK's Biotechnology and Biological Sciences Research Council (BBSRC) to investigate various aspects of the environmental fate of neonicotinoids, and what harm they might be doing. Finally, I had some funds to study these pesticides. To do so, I recruited two postdoctoral scientists, the eternally cheerful and dynamic Beth Nicholls from the West Country, and the quiet, thoughtful and meticulous Cristina Botías from Spain. Between them, they unravelled many of the details about where neonicotinoids go in the environment, and what they do when they get there.

Cristina focused on wildflowers, spending countless hours collecting pollen and nectar by hand from different flowers growing in the field margins and hedgerows of local arable farms in Sussex. We wanted to know if the contamination of dandelions in the US was a fluke, or representative of a more general pattern. Cristina also collected hundreds of soil samples from the field margins where these wildflowers were growing. This was incredibly fiddly work, for the nectar had to be collected by carefully inserting a tiny glass tube into the nectary of each flower, whereupon the nectar was drawn up by 'capillary action'.* Each flower provides just a few thousandths of a millilitre of nectar, and so she had to sample many hundreds of individual flowers to get large enough samples for chemical

---

* Capillary action is the tendency for liquids to flow into narrow spaces, even if uphill, drawn by adhesive forces. Capillary action explains why tissues suck up spilled liquids, and why liquid wax flows up the wick of a candle.

analysis. She also collected pollen by bringing back armfuls of flowers, drying them, then carefully brushing the pollen grains into a test tube. Doing such finicky work would have been difficult enough, but it was made much harder when Cristina developed a pollen allergy, and henceforth had to do all of her field work wearing a respirator, her eyes swollen and red.

The results of Cristina's painstaking sampling were worrying. The soil from the field margins, and the pollen and nectar samples from wildflowers, all frequently contained neonicotinoids that were supposed to only be in the crop. Poppies, brambles, hogweed, violets, forget-me-nots, St John's wort, thistles, all contained the insecticides. The concentrations found were very variable, but sometimes they were much *higher* than those found in a treated oilseed rape crop. For example, some pollen samples from hogweed and poppies contained more than ten times the concentrations we had used in our bumblebee experiment in Scotland (concentrations that the pesticide industry had claimed were unrealistically high). This was initially puzzling, but on reflection was perhaps to be expected. If neonicotinoids were accumulating in the crop soils and in the soil water we might expect them to seep into the field margins. We already knew that crops vary greatly in how much neonicotinoid they suck up through their roots, and so we would expect different wildflowers to also vary. Perhaps poppies and hogweed are just really good at drawing these compounds up from the soil?

Whatever the explanation, it was abundantly clear that the 2013 EU-wide ban on use of neonicotinoids on flowering crops was not going to prevent bees from being exposed to them. Total neonicotinoid use in the UK actually rose after the 2013 ban, owing to a large increase in their use on cereals such as wheat, crops which do not attract bees. If the neonicotinoids applied to cereal crops were getting into wildflowers, then bees and all other insects that visit flowers were still very much at risk. Visiting wildflowers to collect food would seem to have become like a game of Russian roulette for bees, with some free of pesticides but others containing very high doses, and no way for the bees to tell which was which.*

Thankfully, there is a much easier way of collecting nectar and pollen samples from across a wide area than doing it by hand. As Cristina was finding, humans are hopelessly inefficient at doing so. In contrast, bees are masters at extracting floral rewards: they have been collecting pollen and nectar for 120 million years. Their eyes and antennae are finely attuned to the colours and scents of flowers, while their bodies have evolved to help them efficiently gather and carry nectar and pollen. Their abdomen contains an inflatable 'honey stomach', capable of accommodating their own body weight of nectar. Their body is covered in branched hairs which trap pollen, while their legs are equipped with combs with which they deftly brush the pollen into baskets on their hind legs.†

Consequently, bees can be a powerful and efficient tool for detecting environmental contamination of many types, not just by pesticides. A bee colony sends out hundreds or thousands of workers across the landscape, commonly up to a mile or two from their nest, and the bees return with pollen and nectar samples that they have diligently collected from thousands of flowers, which scientists can then steal for their own purposes. That's where Beth, my other postdoc, came in, having drawn the long straw in this research program, for her project involved placing both bumblebee and honeybee colonies out into the landscape and then sampling their nectar stores, and the pollen they were carrying in their pollen baskets.

---

* Research done in Geri Wright's lab in Newcastle found that bees were physiologically incapable of tasting or smelling neonicotinoids, yet, given a choice of feeders containing sugar solution with or without neonicotinoids, they somehow preferred to drink from the poisoned one. This was interpreted by some as evidence that bees can become addicted to the pesticide, in much the same way as a smoker is addicted to nicotine.
† Pollen baskets are found in honeybees and bumblebees, but some other bees carry pollen in other ways. Yellow-faced bees, for example, swallow pollen and then regurgitate it in their nests, while mason and leafcutter bees carry pollen on their hairy tummies.

There are pros and cons to the two different approaches adopted by Cristina and Beth. Sampling food stores from bee colonies is vastly easier than collecting pollen and nectar from flowers, but since we did not know exactly where the bees had been foraging, it was impossible to know exactly where any pesticide residues the food contained had come from. For the pollen we could roughly identify what type of plant it was from peering at the pollen grains under a microscope (pollen grains differ in shape and size between plant species), but again we had no idea where the plant that produced the pollen might be in the landscape. On the other hand, since we were interested in what effects pesticides might have on bees, this approach told us exactly what concentrations of pesticides free-flying bees were being exposed to in the real world.

Via Beth and Cristina's combined efforts, we built up a pretty good picture of the environmental fate of neonicotinoids. Clearly, much of them was going into the soil and remaining there for years. Cristina's soil samples often contained imidacloprid, the first neonicotinoid on the market – but by the time she took her samples, our farmers had all stopped using it several years earlier (imidacloprid was superseded by two newer neonicotinoids, clothianidin and thiamethoxam).* The neonicotinoids were spreading into field margins, and being sucked up by wildflowers and hedgerow vegetation, so that the hedgerows we so value for the farmland wildlife they support were all permeated with potent insecticides. Similarly, flower strips planted along the edges of fields to provide food for bees were also contaminated. In fact, wherever bee colonies were placed in the Sussex landscape, and regardless of whether they were bumblebees or honeybees, both their pollen and nectar stores usually contained neonicotinoids. Often the levels we found in colony food stores were far higher than

---

* Pesticides almost invariably have unpronounceable and hard-to-remember names. Matt Sharlow, CEO of the insect conservation charity Buglife, has a theory that this is a deliberate ploy to discourage public debate about them.

those used in experiments, which, it had been claimed, were unrealistically high. For example, in our Scottish study, which found an 85 per cent drop in numbers of queens produced by bumblebee nests exposed to neonicotinoids, we had given them pollen spiked with 6 parts per billion of the pesticide. Industry spokespeople had argued that 6 parts per billion was an unrealistically high concentration, yet the pollen stores we analysed from bumblebee nests placed in Sussex farmland commonly contained more than 30 parts per billion – clearly more than enough to be doing them harm.

Beth and Cristina also sorted out the pollen balls brought back on the legs of honeybee workers according to what plant they were from, and analysed the different plant species separately to see what pesticides were in them. This was made easier by the fact that on a foraging trip individual honeybees tend to collect pollen from only one plant species, so few of them were carrying mixed loads. Even during the flowering of oilseed rape crops nearby, in April and May, most of the pollen came from wildflowers; hawthorn was the favourite at this time. The data were collected just before the 2013 moratorium on use of neonicotinoids on flowering crops came into effect, but Cristina calculated that only 3 per cent of the neonicotinoid residues coming into the colony via pollen were from crops. Shockingly, 97 per cent of them came in with wildflower pollen.

When Cristina's results were published in 2015, the EU was once again remarkably proactive in its response. The 2013 ban was intended to protect bees from exposure to these chemicals, but clearly preventing their use on flowering crops was not enough. In 2016 the European Commission asked the European Food Standards Agency (EFSA) to review the new evidence and report back, once again focusing on the likely harm to bees. This took well over a year, but when the report arrived in February 2018 it was quite blunt in its conclusions: almost all uses of neonicotinoids pose a risk to bees. In late 2018 all outdoor uses of the three main neonicotinoids were therefore banned throughout Europe.

Our funding did not extend to looking at contamination of aquatic habitats by neonicotinoids seeping in from farmland, but thankfully scientists elsewhere were on the case. Tessa van Dijk and Jeroen van der Sluijs from the University of Utrecht set the ball rolling when they obtained government-collected data on pollution levels in freshwater in the Netherlands, which described alarming levels of neonicotinoids in streams, rivers and lakes. In the most polluted areas concentrations up to a staggering 320 parts per billion were found – such high figures that the stream water itself could have been used as an insecticide. Meanwhile, in Canada, Christy Morrissey from the University of Saskatchewan found near-ubiquitous contamination of Canadian lakes and wetlands with neonicotinoids. When these two studies then sparked scientists worldwide to start looking, it soon became clear that lakes and rivers across the world, from Portugal to California to Vietnam, are often chronically contaminated with these chemicals.

Of all the places studied so far, the Netherlands seems to have the highest levels of neonicotinoid pollution in its fresh water, with levels elsewhere usually below one part per billion. This may not sound like much, but unfortunately it is still enough to kill aquatic insects. In particular, mayflies, caddisflies and some true flies appear to be most sensitive, so that EU regulators have calculated a guideline 'safe' level for the neonicotinoid imidacloprid as 8.3 parts per trillion in fresh water. In a global survey, Christy Morrissey found not only that that this level was exceeded in three-quarters of all samples, but also that up to six different neonicotinoids were found in some samples, and that overall, levels of contamination around the world are increasing year on year.

Not surprisingly, freshwater systems with higher levels of insecticide tend to have less invertebrate life. Healthy streams and lakes can teem with insect life, providing food for fish, birds and bats. In the Netherlands, a wide range of crustaceans and aquatic insects were found to be less common in the more heavily polluted streams, and also the rates of declines of

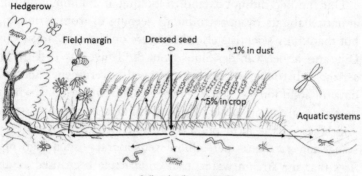

Hedgerow

Field margin   Dressed seed

~1% in dust

~5% in crop

Aquatic systems

Soil and soil water ~94%

The environmental fate of neonicotinoid insecticides used as seed coatings: On average only about 5 per cent of the pesticide goes where it is intended to go – into the crop – a figure that was calculated by manufacturer Bayer's own scientists (see Sur and Stork, 2003, in Further Reading). Most of it ends up in the soil and soil water, where it can build up over time if used repeatedly. From the soil the chemicals can be absorbed by the roots of wildflowers and hedgerow plants, spreading to their leaves and flowers, or they can leach into streams. There is also a fundamental problem with this mode of application, since it is necessarily prophylactic: the farmer cannot know whether the crop will be attacked by pests before he has sown it. Prophylactic use of pesticides is contrary to all the principles of 'integrated pest management' (IPM), an approach that seeks to minimise pesticide use by only applying them when absolutely necessary, and which is regarded by most agricultural scientists as the optimal strategy for pest management. Under IPM, a host of non-chemical techniques for pest management are deployed, such as encouraging natural enemies, using resistant crops, and long crop rotations. Only if these fail, and a significant pest population is detected, does the farmer resort to pesticides.

insect-eating birds in these areas are faster. However, perhaps the most dramatic evidence for an impact of neonicotinoids on freshwater life comes from Lake Shinji in Japan, the study mentioned earlier. Populations of invertebrates here have been closely monitored in the lake for many years, as they support an important eel and smelt fishery. When neonicotinoids were introduced in the surrounding paddy fields they were found to be contaminating the streams that feed into the lake, and populations of insects, crustaceans and other small animals (often collectively known as zooplankton) in the lake immediately collapsed. Apologists for pesticides would argue that this was just coincidence, and that something else may have happened on that precise date that devastated the zooplankton – perhaps there was another pollutant also used for the first time that year, or some devastating zooplankton plague. Such things are of course possible, but ask yourself what *you* think is the most likely explanation.

In Japan, no action was taken, and Lake Shinji fishery yields remained low for at least twenty years (I can find no published data since 2014). In contrast, from a European perspective, the neonicotinoid saga is in many ways an encouraging story. Scientific evidence accumulated, was assessed by regulators, and was acted upon swiftly by government. However, it was a shame that such a terrible mistake was made in the first place, allowing a group of pesticides on to the market for twenty-five years before it became clear how harmful they were to the environment. The

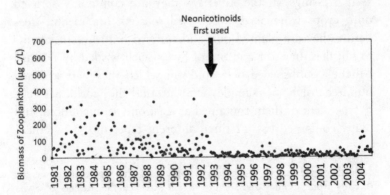

Impact of neonicotinoid pollution on lake invertebrates: Populations of zooplankton in Lake Shinji, Japan, fell dramatically after the introduction of neonicotinoid use on the surrounding rice paddies in 1993 (from Yamamuro et al., 2019).

European Food Safety Authority (EFSA)'s assessments in 2018 of the risks posed by these chemicals examined the evidence for sublethal effects, and their effects on wild bees (not just honeybees), but there is still no requirement for new pesticides coming to market to be assessed in this way. And although the EFSA worked with scientists from across Europe to develop a protocol for more rigorous assessment of new pesticides, intensive lobbying of politicians by the industry has so far succeeded in preventing it from being implemented. Essentially there is nothing to prevent exactly the same mistakes being made again with new chemicals. New insecticides have appeared on the market in recent years, with the usual unpronounceable names: flupyradifurone, sulfoxaflor, cynatraniliprole, and more. Most are potent neurotoxins, with properties similar to neonicotinoids. Indeed, some of them arguably *are* neonicotinoids. One wonders whether, twenty years from now, these compounds too will be banned, once sufficient evidence accumulates.

Yet although Europe has been proactive in banning neonicotinoids, first on flowering crops and then on all crops, the tragedy is that they remain the insecticide of choice throughout the rest of the world. Aside from their near-ubiquitous use as seed dressings, in the Americas they are commonly sprayed onto crops from aeroplanes, used to soak ornamental trees before they are planted out, and sprayed onto livestock units to kill flies. In 2017 a group of Swiss scientists led by Edward Mitchell published a new study in which they had screened hundreds of honey samples from around the world. Seventy-five per cent of them contained at least one neonicotinoid, and many contained two or three different types mixed together.*

---

* Further evidence of the pervasive nature of these pesticides emerged recently from another Swiss study, which screened the feathers of Swiss house sparrows for neonicotinoids. Of a large sample of several hundred birds, which included birds living on organic farms, 100 per cent of them contained at least one neonicotinoid. Meanwhile, in the USA researchers have found that plausible doses of neonicotinoids reduce the body weight and mess up the navigational abilities of migrating white-crowned sparrows.

Even bees on remote islands, such as Curaçao in the Caribbean and Tahiti in the Pacific, had these toxins in their food stores.

This is worth dwelling on for a moment. Honeybees are vitally important insects; that three-quarters of them have highly potent insecticides in their food should be a cause of the deepest concern. Of course, this poses a very serious threat to honeybees, given the range of lethal and sublethal effects associated with these chemicals, but more broadly it indicates a global threat to all pollinating insects. Honeybees are generalists: they will have gathered this honey from an enormous range of different plant species. So if honeybees are being exposed to neonicotinoids, then so too are bumblebees, solitary bees, butterflies, moths, beetles, wasps and so on. It now seems likely that a large proportion of all the world's insect species are being chronically exposed to chemicals specifically designed to kill insects.

Even in Europe the problem is not yet completely solved. Farmers can apply to their national government for a derogation, a legal term meaning they are granted temporary exemption from the ban, so they have special permission to use neonicotinoids on their crops. They have to argue that it is an emergency use, and that no suitable alternative method of pest control is available, but many EU governments seem willing to dish out these derogations without examining the plausibility of the argument that they are necessary. In 2017, for instance, thirteen out of twenty-eight EU countries granted derogations to farmers allowing them to use the banned neonicotinoids on flowering crops, while one of the first actions of the UK government post-Brexit in January 2021 was to grant a derogation for their use on sugar beet, despite fierce protests from environmental groups.

In addition, the EU ban does not extend to veterinary uses. While farmers are no longer allowed to use neonicotinoids on their crops (unless they get a derogation), the public can still buy them as flea treatments for their pets. Advocate and Advantage are among the most popular brands of flea treatments, and both have imidacloprid as the active ingredient. A competing product, Frontline, has fipronil as its active ingredient, another neurotoxic insecticide with very similar properties to neonicotinoids. You might think that use on pets would be trivial compared to that on crops, and certainly the amounts involved are smaller, but nonetheless the doses used are high. These flea treatments are intended for prophylactic use, to be dripped onto the back of the neck of your dog or cat every month, rendering the animal toxic to blood-sucking insects. The dose recommended for a medium-sized dog each month is enough to kill sixty million honeybees, and with about ten million dogs and eleven million cats in the UK alone, it seems likely that several tons each of imidacloprid and fipronil are going onto our pets each year.

Of course, this might not be an issue, since bees don't usually feed on dogs or cats, but remember that these are persistent and water-soluble chemicals. If a dog jumps into a pond or stream, or goes out in the rain, the toxin is surely likely to wash off, delivering a sizeable dose of insecticide into the environment. My PhD student, Rosemary Perkins, recently obtained and analysed data from the Environment Agency on levels of imidacloprid and fipronil in twenty English rivers. Her results were highly concerning, for nineteen of the rivers were contaminated with imidacloprid, and all twenty with fipronil and various toxic breakdown products of fipronil. Worse, in the majority of rivers the levels found far exceeded safe limits for aquatic insects. Tellingly, levels of contamination with both chemicals were higher in river samples taken downstream of the outflow from sewage works. A US study found that bathing dogs released most of the flea treatment into the bath water, so it seems likely that this is the source of contamination of English rivers, particularly since fipronil was never licensed for agricultural use in the UK.

During Beth and Cristina's studies of the pesticides in food collected by bee colonies, we had placed some bumblebee nests in urban areas. We were curious to see how pesticide exposure differed between town and country. In the UK, pesticide use by farmers is carefully monitored by government agencies, but there seems to be no monitoring of pesticide use by gardeners, local authorities, or by pet owners. Overall, concentrations of pesticides in food stores of bumblebee nests in urban areas tended to be lower than those in nests in the countryside, but the pesticides we found were different. In the countryside the most common neonicotinoids in bee food were clothianidin and thiamethoxam, newer compounds that replaced imidacloprid several years ago for farm use. By contrast, the main neonicotinoid in the food stores of town bees was imidacloprid. We still don't know for sure where it is coming from. Imidacloprid used to be the main ingredient of many bug sprays sold for garden use, and it could be that some gardeners were still using old bottles bought years earlier. It could also be that imidacloprid is still hanging around in soil and plants from applications made years ago. Or it could be that it was coming from dogs and cats, washing off their fur during rainfall and contaminating gardens. It's most likely to have been all three, though it would be nice to know the relative contributions, since two of these sources will eventually cease, while there is no sign that the neonicotinoid flea treatments will be withdrawn.

I have concentrated thus far on neonicotinoids, the most notorious pesticides still in use, but they are just the most prominent part of a much larger issue. Much as scientists made the mistake of feeling that the pesticide issue had been dealt with in the battles over DDT and its ilk in the 1960s to 1980s, so it would be a major mistake to assume that neonicotinoids are the only pesticides that pose a risk to insects, or to the environment in general. Some scientists and campaigners have perhaps developed tunnel vision, focusing all their efforts on this one subject, and losing sight of the bigger picture. In our Sussex studies of the pesticides in the pollen and nectar stores of bee nests we also screened for fungicides, and these were usually more common than insecticides. In pollen collected by bumblebees we found a minimum of three different pesticides, and sometimes as many as ten, all mixed up together. Other researchers around the world have screened honey and pollen samples for all types of pesticides and, no matter what the geographic location of the hive, the bees' food almost invariably contains a complex cocktail of insecticides, fungicides and herbicides. One hundred and sixty different pesticides have been found in bee food stores, including 83 different types of insecticide, 40 fungicides, 27 herbicides and 10 acaricides (the latter a treatment intended to kill mites).

Of course, one might reasonably expect that pesticides such as herbicides and fungicides would not be much of a problem for insects, since they are not intended to be toxic to them. Herbicides are mainly applied to kill weeds within or near the crop, while fungicides are applied to crops to control fungal diseases such as mildews, rusts and blights, which left unchecked can do great harm, particularly in wet and humid weather. Farmers often assume that fungicides and herbicides are harmless to bees, and so will spray them onto flowering crops during the day, when bees are active (something they would generally avoid doing with insecticides). Yet evidence is emerging that fungicides are indeed harmful to insects. For example, a large-scale study of patterns of decline in North American bumblebees found that the best predictor of decline overall was the use of fungicides, not insecticides or herbicides. This study also found that use of one particular fungicide called chlorothalonil was strongly correlated with incidence of a disease, *Nosema bombi*, a cause of sometimes lethal diarrhoea in bumblebees. Independently, other researchers have found that honeybees exposed to this chemical are more susceptible to the closely related disease, *Nosema ceranae*, and also that exposing bumblebee colonies to chlorothalonil at realistic concentrations – ones we might expect bees living on farmland to receive – was enough to reduce colony growth significantly. It isn't certain how this fungicide harms bees, but one theory is that it kills beneficial gut microbes, and it is this that makes the bees more susceptible to diseases.*

The fungicide chlorothalonil has been in use since 1964, and is one of the most widely used pesticides in the world. Until 2018 it was the single most-used pesticide in the UK, and it is very commonly found in honey samples. Its harmful effects on bees were not picked up during its initial registration, and seem to have escaped notice for more than fifty years. In 2019 this chemical was banned by the EU, primarily over concerns that it was contaminating groundwater – and hence getting into rivers and drinking water – rather than anything to do with bees. As with neonicotinoids, the rest of the world continues to use it without restriction. One has to wonder how many other chemicals of the hundreds that are currently in use will eventually turn out to be harmful to something, be it bees, humans or great crested newts. Our regulators reassure us that they are safe until they tell us that they are not. How can we have any faith in a system that has misled us so many times in the past?

While some fungicides seem to be directly harmful to bees, others have more subtle effects. For example, a class of chemicals snappily known as ergosterol biosynthesis-inhibiting (EBI) fungicides are now known to act synergistically with insecticides. The fungicide blocks the detoxification mechanism of the bee, which is not a problem if the bee is not exposed to a poison. However, if the bee is simultaneously exposed to an insecticide, its ability to cope with it greatly decreases. Some fungicide-insecticide mixtures can be up to 1,000 times more toxic to bees than the insecticide alone. Such unexpected interactions will never be picked up by the regulatory tests for new pesticides, since each chemical is only ever tested in isolation.

---

* Just like humans, and probably almost all animals, bees have complex communities of symbiotic bacteria living in their gut, and disruption of these communities can profoundly affect their health.

Yet, of course, we know that in the real world bees and other insects are almost never exposed to just a single chemical at a time. In reality, from the day they hatch from their egg onwards they are likely to be exposed to a complex mixture of man-made chemicals, and also to both naturally occurring and novel pathogens and parasites (see chapter 10), the interactions being far beyond our ability to predict or understand. As a result, even DEFRA's chief scientist Ian Boyd recently admitted that it is not currently possible to predict the environmental repercussions of landscape-scale use of large quantities of multiple pesticides. Yet we carry on doing it anyway.

Let us return to the claim that pesticide use today is safer for the environment than in the bad old days of DDT and its ilk. After all, advocates of pesticides are always keen to point out that the total weight of pesticides applied to the landscape has fallen. This is certainly true in the UK: during the period from 1990 to 2015 the weight of pesticide applied by farmers fell from 34.4 thousand tons to 17.8 thousand tons, a drop of 48 per cent (globally, the total weight of pesticides applied in 2015 was about 400,000 tons, so the UK accounts for about 4 per cent of global use). These figures, I should explain, are the weight of the 'active ingredient', meaning the actual toxin, which is often formulated with a bunch of other chemicals[*] in a much larger volume of water or other solvent.

However, the apparent decline in pesticide use is deceptive. All else being equal, applying less pesticide must be a good

---

[*] These so called 'inert' ingredients are not subject to the same regulatory tests as the 'active' ingredient, but recent evidence suggests that the combined effects of the mixture can be much more toxic than the 'active' ingredient alone. The pesticide formulations containing the mix of active and inert ingredients that are sold to farmers are given dramatic, powerful or sometimes just plain strange names. Amongst the best known are formulations containing glyphosate, which are usually known as Roundup. Various insecticide formulations include Shadow, Cruiser, Mavrik, Advocate and even Gandalf – the latter now being a pyrethroid insecticide. I doubt that J. R. R. Tolkein would approve.

thing, but all else is far from equal. A feature of these newer compounds is that, as time has gone on, they have tended to become much, much more toxic than what went before. In 1945, DDT was typically applied at a rate of about 2,000g per ha. More modern insecticides such as aldicarb, pyrethroids and neonicotinoids are applied at about 100, 50 and 10g per ha, respectively, because they are much more poisonous to insects – and that means both the good insects and the 'bad' ones. A quick back-of-an-envelope calculation suggests that the net effect of switching to using chemicals that are used in smaller amounts but are much more toxic might pose a greater risk to insects. Neonicotinoids and fipronil are about 7,000 times more toxic to bees than DDT, so swapping 2kg of DDT (which could kill about 74 million honeybees) for 10g of a neonico-tinoid (which could kill 2.5 billion honeybees) is not a step in the right direction, at least from the point of view of the bee.

I decided to look into this in more detail and, with the help of a succession of keen students at Sussex University, I set about working out how the potential threat posed to bees by pesticide use has changed over time in the UK. To DEFRA's credit, figures on pesticide use on crops by farmers in the UK are freely available on a government website, https://secure.fera.defra.gov.uk/pusstats/, and are updated every year. For each of the 300 or so pesticides that are regularly used in the UK, we extracted the weight used each year and calculated how many honeybees it could, theoretically, kill. For each year, we summed the total 'potential bee mortality' from all the different pesticides, and looked to see how it changed over time. Of course, I should stress that this is a highly unlikely worst-case scenario, in which all of the pesticides applied by farmers were consumed by bees. Nonetheless, the scale of the vertical axis in the diagram is alarming – the pesticides applied to the UK could kill every one of the approximately three trillion honey-bees on the planet several thousand times over – so it is a good thing that most of them are never consumed by bees. It is the pattern that I think is important: since 1990 the number of

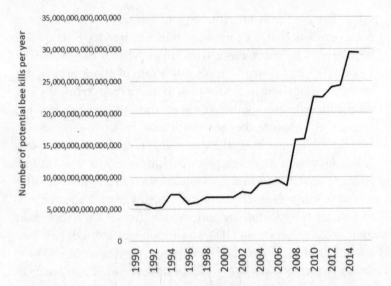

Changing 'toxic load' over time: The chart shows the potential number of honeybees that could be killed by the pesticides applied to UK crops each year, in the unlikely event that all of them were fed to honeybees. The number has increased six-fold since 1990, as newer, more toxic insecticides have been adopted by farmers. From https://peerj.com/articles/5255/. Note that this does not include the considerable volume of ivermectins fed to cattle, a class of pesticides that are highly toxic to insects and present in large quantities in livestock dung, contaminating soils.

potential bee deaths has risen six-fold. From a bee's perspective, farmland has become a much more dangerous place.

Of course, the pesticides used by farmers are not intended to kill honeybees, or bumblebees, or any other beneficial or harmless insect. They are targeted as best they can be against pest insects such as aphids, whitefly and herbivorous caterpillars, for example by spraying in the evening when most bees will be asleep, and so the vast majority of the weight applied does not go near a bee. Yet one cannot broadcast 17,000 tons of poison across the landscape every year without doing unintentional harm. Insecticides kill *all* insects, not just the ones that they are aimed at, whatever the doublethinkers who manufacture agrochemicals may ask you to believe. It is little wonder that our wildlife is in trouble, for any plant or animal living in arable farmland has to cope with being sprayed over and over again each year. Pesticides sprayed from tractor booms (or from crop duster aircraft, as commonly practised in the Americas) can drift into hedges and beyond. Those used as seed dressings can accumulate in the soils and seep into water courses. With more than seventeen pesticide applications to every arable field each year, it is inevitable that much of our countryside is contaminated with pesticides.

Unfortunately, the pesticides used on crops are only part of the story. Livestock farmers, for instance, dose their livestock regularly with ivermectins, which protect them against intestinal worms and insect parasites. Usually taken orally, most of the ivermectins pass through the animal and into their dung, where the chemicals can persist for months, rendering what ought to be a feast for dung beetles and flies into a toxic hazard. At the same time, local authorities spray our parks and pavements, while homeowners blast their gardens with poisons bought from the supermarket or DIY store and drip them on their dogs and cats. Unbelievably, it is even possible to buy Woodlice Killer. Indeed, the gardening section of a national newspaper recently promoted 'controlling' woodlice with Vitax Ltd Nippon Wood Lice Killer if they become too numerous in compost heaps. For those inclined to follow the advice, this product contains a general-purpose pyrethroid insecticide, and is available from Amazon online, or from garden centres and DIY stores. It is advertised as being suitable for outdoor and indoor use, with the claim that it is also effective against earwigs and silverfish. Quite why one would wish to kill any of these organisms is unclear. Woodlice are benign creatures, doing a fantastic job in the compost heap of chewing up woody material and helping it become dark, rich compost. They particularly thrive in damp wood piles, quietly recycling the nutrients from the wood and eventually making them available to your garden plants. They are food for birds and small mammals. These are beneficial creatures, and should be celebrated, not persecuted and poisoned in some misguided psychotic urge to kill anything that dares to thrive. Woodlice cannot become 'too numerous' in compost heaps, as they are actually helping to break down the compost; the more the merrier. If woodlice and silverfish regularly turn up in numbers in your house, you have a damp problem. It is probably best to tackle that, rather than treating the symptom by dousing your home with pesticide.

The big yet invisible picture is that our soils, rivers, lakes, hedgerows, gardens and parks are all now contaminated with mixtures of man-made toxins. It is sometimes said that humanity is at war with nature, but the word 'war' implies a two-way conflict. Our chemical onslaught on nature is more akin to genocide. It is small wonder that our wildlife is in decline.

The Earwig's Second Penis

It is a little-known fact that the males of many species of earwigs have two penises, a feature they share with snakes.

Researchers in Japan have found that earwigs are predominantly right-'handed', with 90 per cent of males using their right penis for mating. However, if the right penis is chopped off (some scientists get up to strange things), the earwig simply uses the left one, which seems to work just as well. By strange coincidence, a colloquial name for earwigs in some regions of Japan, *chimpo-kiri*, means 'penis-cutter', perhaps because earwigs were often found around old-fashioned outdoor toilets.

Why, though, would an earwig need two penises? If experimentally disturbed during mating, it is common for the male to snap off the penis in use, leaving it inside the female, so that he can make a swift escape. The abandoned penis serves to plug up the female, reducing the likelihood of her mating again, and thereby increasing the chances of siring her offspring. The second penis serves as a back-up, enabling the male to mate again. Scientists have not investigated whether they snap off their second penis so readily.

Interestingly, many male spiders adopt a similar strategy. Males transfer sperm to females via one of a pair of palps, and in some species the palp is always snapped off and left behind in the female, meaning that the male can only ever mate twice. The palp continues to pump sperm into the female after its owner has departed. In spiders, this mechanism may have evolved because females commonly eat their lover if he hangs around for too long.

8

Weed Control

The most common pesticides used on most farms are herbicides, chemicals that farmers find helpful for removing non-crop plant species – usually called weeds – that might otherwise compete with the crop, so reducing the yield. They are also often used to deliberately kill off a mature crop, such as wheat or cotton, ensuring that it dies and dries uniformly. This makes it easier to harvest and allows the farmer some control over when he will harvest so that he can co-ordinate it with other farm activities, though with the unfortunate downside that there will be herbicide in the harvested crop. The best-known and perhaps most notorious herbicide is glyphosate, a chemical usually sold in a formulation known as Roundup. Glyphosate is the most widely used pesticide in the world, with use in UK farming increasing year on year to over 2,000 tons in 2016. This figure does not include use by local authorities or in domestic gardens, both of which must be considerable, given how frequently one sees the tell-tale yellow, scorched vegetation alongside paths, pavements and road verges, but these uses are not monitored by government or anybody else.

Glyphosate is a general-purpose herbicide, killing any plant it touches. It is systemic, which means that it spreads through the tissues of the plant to kill the roots. I hate to admit this now, but I once used to use it quite a lot in my garden, as I believed the manufacturers when they claimed that it was non-toxic

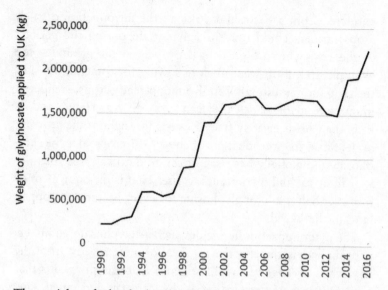

The weight of the herbicide glyphosate used by UK farmers: Glyphosate is most commonly sold as the formulation Roundup, and is the single most popular pesticide in the world, with use increasing year on year. The figures shown do not include domestic use, or use by local authorities. Data are from DEFRA's Pusstats website, an open-access database which reports annual use of pesticides in UK farming.

to wildlife and broke down very quickly in the environment. I used to be very naïve. I found it useful for knocking back weeds that are hard to get rid of by digging, such as nettles, bindweed, couch grass and brambles, although even when treated with glyphosate these tough plants always seemed to sprout back after a while. It was also very handy for keeping paths and driveways clear, saving the effort of dibbing out weeds sprouting from the cracks with a hand trowel. You will soon understand why I do not use it any more.

The 2,000 tons used in the UK might sound a lot, but by international standards it is a trifling amount. Outside Europe, much glyphosate is used in association with 'Roundup-ready' genetically modified crops, which have been rendered immune to the effects of the herbicide by giving them an extra gene, extracted from a bacterium. Prior to the introduction of such crops farmers could only use glyphosate during the periods of the year when no growing crop was in the ground – for example, to kill off weeds before sowing (or sometimes to kill the crop for harvesting). With Roundup-ready crops – rendered immune to glyphosate – they can spray all year round, and so keep their fields entirely free of weeds throughout the growing cycle. Since the introduction of these GM crops in 1996, glyphosate use globally has risen fifteen-fold, to 825,000 tons per year in 2014, and it continues to rise. That is the equivalent of just over half a kilo of glyphosate for every hectare of cropped land in the world.

The main impact of herbicides such as glyphosate on insects is probably via the effective removal of most 'weeds' from the landscape. Weeds, we should remember, are simply plants that are in the wrong place, so far as the farmer is concerned. One man's weed is another man's wildflower. Most arable farmers and gardeners in Europe and North America would agree that couch grass (*Elymus repens*) is a troublesome weed – we are shallow creatures, and do not value it because it does not produce pretty flowers – yet it is an important forage plant for livestock, the seeds are popular with goldfinches, and the leaves are eaten by caterpillars of the Essex skipper butterfly. Thistles would ordinarily be regarded as weeds, but their flowers are much loved by bees and butterflies, the seeds are a favourite with finches, and dozens of herbivorous insects eat the leaves or burrow in the stem or flower head, in turn providing food for insectivorous birds.* And then there are poppies, cornflowers, corn marigold, corn cockle and many other beautiful flowers that many of us love to see, but which tend to turn up in arable fields, can compete with the crop, and so are commonly regarded as weeds.

---

* I'm happy to report that I have creeping, marsh, spear and woolly thistle in my garden.

Plants are, of course, the basis of almost every food chain, and by developing methods of farming that almost entirely eradicate weeds from arable fields, such that crops are often close to pure monocultures, we have made much of our landscape inhospitable to most forms of life. For example, there are seventy-three different species of milkweed that are native to the USA, thirty of which are used as caterpillar food-plants by monarch butterflies (they will eat nothing else). It has been estimated that the abundance of milkweed in the landscape of the American Midwest fell by 58 percent in just eleven years from 1999 to 2010, and this is likely to be a major driver in declines of the monarch butterfly. Although the herbicide glyphosate is the prime suspect behind monarch declines, the herbicide dicamba has also been implicated. As with glyphosate, use of this herbicide has recently increased with the introduction of genetically modified, dicamba-tolerant cotton and soybean crops, which allows the farmer to spray the growing crop and wipe out all non-crop plants. Unfortunately, it has recently emerged that dicamba becomes volatile during unusually hot weather – for example in summer heatwaves, which are becoming more common under climate change – and the chemical vaporises and then drifts on the wind far from its target, often killing both crops and wildflowers such as milkweed hundreds of metres downwind.

Insect declines, of course, have knock-on effects for other organisms, such as those that eat them. When I was a boy growing up in rural Shropshire in the UK, the grey partridge was a common farmland bird, but since 1967 it has declined by 92 per cent. Long-term scientific studies performed in Sussex showed that, perhaps surprisingly, herbicides were the main driver of this decline. It was not that the herbicides were poisoning the birds, but rather that they were greatly reducing the number of weeds in crops and field edges, which in turn was reducing the abundance of caterpillars and other herbivorous insects that are the main diet of partridge chicks. Similarly, over the same period cuckoos have declined by 77 per cent.

They are specialist predators of large, hairy caterpillars such as the ridiculously mop-haired, 'woolly bear' caterpillars of the garden tiger moth. Both caterpillars and the adult moths were a common sight thirty years ago, with the orange and black caterpillars crawling speedily about on the ground in search of dandelions and other favoured leaves to eat, while the spectacularly colourful chocolate, cream and scarlet adults would often be seen sitting about in the morning near outside lights. Sadly, this species declined by 89 per cent between 1968 and 2002, thought to be largely the effects of increased tidiness of the countryside enabled by weedkillers.

As with insect declines, the loss of wildflowers from our landscape has gone largely unnoticed. Corn marigolds, corn cockle and cornflower all used to be common, but today they are scarcely ever seen in fields used for crops. Poppies have survived better, probably due to the extraordinary longevity of their seeds in the soil, which can last for many decades. Even so, fields of red poppies are now a rare sight, and I suspect that the seed bank is depleted further every year. A review of evidence from Germany calculated that the number of weed species found on farms fell by between 50 and 90 per cent between 1945 and 1995, depending on the region, with an overall mean of 65 per cent. On average, the number of weed species that could be found per field fell from twenty-four to only seven.

Globally, 571 plant species have gone extinct since records began. The number of plants that have disappeared from the wild is more than twice the number of extinct birds, mammals and amphibians combined. There is also broad agreement that this is a huge underestimate of the true figure, for many more species have not been seen for decades, but have not been declared extinct in case some still survive in some remote corner of the world. Plant extinctions tend to attract much less attention than those of animals, which we tend to find more charismatic; can you name a single one of those 571 extinct plants? Who, for instance, has heard of the Chilean sandalwood, or the Appalachian yellow asphodel, and the St Helena olive, may they rest in peace? Every loss of a plant species is likely to trigger a cascade of further extinctions, for most plants have associated insects or other animals. As with the loss of insects, the decline of plant species deserves far more attention.

There are also growing concerns that herbicides may be directly toxic to insects, and indeed to us. Twenty-seven different herbicides have been found in samples of honey and pollen from honeybee hives, showing that pollinators in general are probably often exposed to them. Of course, herbicides are only supposed to be poisonous to plants, and animals and plants are quite different (sorry for stating the obvious), so one might think that it ought to be possible to find herbicides that aren't harmful to us animals. Nonetheless, we seem to have failed to do so.

Let's return to look in more detail at glyphosate. It is intended to target an enzyme only found in plants and bacteria, and so it ought to have no effect on animals. Yet recent studies on honeybees by Erick Motta at the University of Texas have found that exposure to glyphosate in their diet alters the beneficial gut flora of the bees, making them more susceptible to disease (note that exposure to the fungicide chlorothalonil is also suspected to cause this). On reflection this isn't surprising, since we know that glyphosate is toxic to many bacteria, and it has also become apparent that bees, just like humans, have communities of bacteria living in their guts that have a profound effect on their health and immunity to disease.

Nor is upsetting gut bacteria the only effect that glyphosate seems to have on honeybees. The Argentinian scientist Maria Sol Balbuena used high-tech harmonic radar to track the homing abilities of bees, and found that those given small doses of glyphosate took longer to find their way home when released at an unfamiliar release point, and also took more roundabout routes, compared to control bees (fascinatingly, neonicotinoid insecticides have this same effect on bees). The effect occurred immediately after consumption, and so could not be caused by any impact on the gut flora, which take days or weeks to have effects on host health. Other studies have found that glyphosate impairs the learning of associations between floral odours and rewards, something bees are normally very good at, and vital if they are to efficiently collect lots of pollen and nectar. Both poor navigation and impacts on learning might be explained by some sort of effect on memory retrieval, but we don't yet know how this might work.

Once again, these recent revelations of harmful effects of glyphosate on bees highlight the shortcomings of the current regulatory system, which relies primarily on short-term lab studies of toxicity. Bees with poor navigation and learning, and depleted gut flora, would appear completely normal in a lab setting, leading to the erroneous conclusion that the chemical was harmless to them.

Given that we also have a gut flora, at this point you'd be justified in wondering what impact glyphosate has on us. This has become the subject of great controversy. It is beyond doubt that we are all exposed to glyphosate on a daily basis, not least because glyphosate has proven to be far more persistent than was previously thought, lasting for months in soils and for a year or more in pond sediments. It clearly survives the harvesting and processing of crops, and, in part because of the common practice of spraying wheat crops with glyphosate just before harvest, it is very commonly found in cereal-based foodstuff such as bread, biscuits and breakfast cereals. For example, in the USA concentrations of glyphosate of several hundred parts per billion have been found in Quaker Oats, Nature Valley granola bars and Cheerios cereal, among other everyday products. As these concentrations happen to be higher in many products aimed at children, the US Environmental Protection Agency has calculated that children aged one to two years old are likely to receive doses exceeding what they call 'no significant risk'. This would seem to be a convoluted way of acknowledging that there is a risk to children.

Given the global use of glyphosate, it seems likely that we are all exposed, one way or another. A recent study from

Germany found that over 99 per cent of a sample of 2,000 people had detectable glyphosate in their urine, with children tending to have more than adults. What impacts, if any, does this have on us? Herein lies the controversy. In 2014, a 'meta-analysis' – a pooled analysis of all relevant data sets on the subject – concluded that there was an increased risk of the cancer non-Hodgkin's lymphoma in people who were occupationally exposed to glyphosate. In March 2015 the World Health Organization's International Agency for Research on Cancer (IARC) concluded that glyphosate was 'probably carcinogenic to humans'. They based this on their assessment that there was strong evidence that glyphosate can cause oxidative stress (by using up the body's supply of antioxidants), and is genotoxic (damaging genetic information, causing mutations that can lead to cancer).

Eight months later in November 2015 the European Food Standards Agency (EFSA) published a report in which it concluded that glyphosate was not carcinogenic, directly contradicting the IARC. This was almost immediately followed by the publication of a strong critique of the EFSA report in a paper written by no fewer than ninety-four authors, including many eminent toxicologists and epidemiologists from around the world. Then the following year, in 2016, the US Environmental Protection Agency (EPA) released a report that concurred with EFSA but disagreed with IARC, concluding that glyphosate is 'not likely to be carcinogenic to humans at doses relevant to human health risk assessment'. Since 2016, further scientific reports and reviews of evidence and methods have variously criticised or supported IARC, EPA and EFSA.

For ordinary people this is extraordinarily confusing. Even for scientists such as myself, trained to weigh up scientific evidence, it is hard to know what to conclude. How can groups of scientists and scientific organisations, each with access to essentially the same sets of data, come to such contradictory conclusions? And if the professional toxicologists cannot agree, what are the rest of us supposed to believe?

The American agricultural scientist Charles Benbrook published a detailed comparison of the methods used by IARC and EPA, revealing considerable differences. The IARC assessment that glyphosate could cause cancer was based very largely on peer-reviewed* studies, for instance, while the EPA assessment relied more heavily on studies carried out by Monsanto (the manufacturers of glyphosate) themselves. So the IARC was relying on data checked by independent experts in the field, while the EPA in the United States was using studies supplied by the manufacturers themselves, which had never been subject to peer-review by independent scientists.

Allowing companies to evaluate the safety of their own chemicals remains standard practice around the world, despite the obvious conflict of interest that this creates. To make matters worse, such studies are often never made available for public scrutiny. As Charles Benbrook points out, the findings of such regulatory studies performed by the pesticide industry contrast remarkably with those in the peer-reviewed scientific literature. For example, the EPA report included 104 studies on whether glyphosate was genotoxic (damaging to DNA, causing mutations and cancer), 52 of which were conducted by Monsanto, and 52 taken from the public, peer-reviewed literature. Just one of the 52 studies conducted by Monsanto found genotoxicity (2 per cent), in contrast to 35 of the 52 studies from the publicly accessible scientific literature (67 per cent). You do not need to be a statistician to conclude that something is wrong here. One explanation might be that scientists not working for agrochemical companies might not always bother to publish negative results; showing that something does not

---

* All scientific studies are normally subject to a peer-review process before they are published. The authors of the study submit their work to a journal, and the journal editor then asks at least two independent experts to evaluate the quality of the work, usually anonymously. The system isn't perfect, for errors can still slip through, but in general peer-reviewed papers are considered to be far more reliable than reports and studies that have not been through this process.

cause cancer is much less exciting than showing that it does. They may also find such negative results harder to publish because journal editors like to publish newsworthy papers ('Glyphosate causes cancer' might make the news headlines, but 'Glyphosate seems harmless' is less compelling). Equally, scientists who do work for agrochemical companies might feel under pressure (consciously or subconsciously) to find no harmful effects of their employer's products. One way or another, it is clear that there is a huge discrepancy between the two data sets.

There are other important differences, Charles Benbrook found, between the approaches used by IARC and EPA. The EPA assessment was focused primarily on studies using pure glyphosate, whereas the IARC gave equal weight to the smaller number of studies on the effects of glyphosate formulated with various other compounds in 'glyphosate-based herbicides'. In reality, all exposure of wildlife and humans is to glyphosate-based herbicides, for farmers never apply pure glyphosate. The pesticides sold to farmers are sold under a brand name, such as Roundup, which consists of a formulated mix of the 'active ingredient' – in this case glyphosate – with other chemicals intended to make the herbicide more effective. For example, these might include wetting agents (detergents) that help the product stick to plant leaves, rather than dripping off. In the real world, farmers, bees and butterflies are exposed to Roundup, not to pure glyphosate. As well as sticking to plants better, there is evidence that the formulated products stick to animals better too, and that these extra ingredients can also increase absorption through the skin, and affect how the product is metabolised in the body. Overall, they make the herbicide more toxic than if it were pure glyphosate, and sometimes hundreds of times so. It seems illogical that regulatory tests should focus on pure glyphosate, when clearly what matters is the effect of exposure to the formulated product.

The main breakdown product of glyphosate is aminomethylphosphonic acid (AMPA to its friends), a toxic chemical also widely found in human foodstuffs. The IARC included studies of the toxicity of AMPA, while the EPA did not.

Finally, Charles Benbrook points out that the EPA assessment of likely exposure of people was focused on exposure of the general public via the consumption of contaminated foods, rather than occupational exposure of farmers, groundskeepers or gardeners. These groups are of course likely to receive much higher doses, particularly during accidental spillages or, for example, if using leaky backpack sprayers. With millions of people around the world applying glyphosate on a regular basis, such accidents are inevitable.

Benbrook clearly feels that the IARC report is more reliable than that of the EPA. American juries seem to agree. In August 2018 a Californian jury found unanimously in favour of forty-six-year-old Dewayne Johnson, a school groundskeeper who

A subsistence farmer in Bengal, India, spraying herbicide using home-made equipment. Note the lack of any protective mask, gloves, or even footwear.

had developed non-Hodgkin's lymphoma after years of using glyphosate as part of his job. During training in pesticide use, Dewayne had been told that glyphosate was 'safe enough to drink'. Monsanto were ordered to pay him the staggering sum of $289 million in compensation and punitive damages. The jury's decision included a statement that Monsanto's products were a 'substantial danger' to the public, and the verdict stated that Monsanto had 'acted with malice'. Of course, just because a jury of non-experts agree on something, does not necessarily make it true. The jury may have seen through the smoke clouds to the scientific truth, or they may have been swayed by a natural tendency to find in favour of the little guy, and to be suspicious of big corporations.

However, in March 2019 a second Californian jury found against Monsanto, this time in favour of Edwin Hardeman, who had argued that he had contracted non-Hodgkin's lymphoma after thirty years of using glyphosate on his garden and in rental properties he owned. He was awarded $80 million, with the court ruling that Roundup was defectively designed, that Monsanto failed to warn of the cancer risk it posed, and that the company had been negligent. 'Mr. Hardeman is pleased,' stated Hardeman's lawyers Aimee Wagstaff and Jennifer Moor after the trial,

> that the jury unanimously held that Monsanto is responsible for causing his non-Hodgkin's lymphoma. As demonstrated throughout trial, since Roundup's inception over forty years ago, Monsanto refuses to act responsibly. It is clear from Monsanto's actions that it does not care whether Roundup causes cancer, focusing instead on manipulating public opinion and undermining anyone who raises genuine and legitimate concerns about Roundup.

Almost simultaneous with the announcement of this verdict was the publication of another academic study, this time by Luoping Zhang and colleagues from the University of California in Berkeley. Their new 'meta-analysis' concluded that people who are routinely exposed to glyphosate via their occupation (farmers, groundskeepers and so on) have a 41 per cent higher risk of developing non-Hodgkin's lymphoma.

In a third Californian trial in May 2019 a married couple, Alva and Alberta Pilliod, who both contracted non-Hodgkin's lymphoma after many years of glyphosate use, were awarded $2 billion, though it looks as if this phenomenal sum will probably be reduced on appeal. Monsanto is now said to face more than 13,400 more lawsuits brought by people suffering from cancer who blame glyphosate for their disease. In 2018 Bayer purchased Monsanto for $63 billion, just before the verdict on Dewayne Johnson's case was announced. It is a move the German company must surely be regretting, for Bayer's share price has since dropped by about 40 billion euros.

Nonetheless, Monsanto still continues loudly to protest its innocence. In 2016 the company set aside a budget of $17 million specifically for championing the safety of glyphosate and challenging the IARC finding that glyphosate was a carcinogen.

If I seem to have rambled a way off topic here, I do so because this story nicely illustrated the difficulties we all face in trying to weigh up scientific evidence and assign cause and effect, whether we are a juror, a scientist, or simply a layperson trying to decide whether to spray glyphosate on the weeds on our drive or get down on our hands and knees to dig them up by hand. It is particularly difficult when corporate interests are involved and substantial sums are being spent on trying to promote a particular view. There are obvious parallels with the neonicotinoid debate, and previously with the long-running saga of tobacco and human health.

For my part, I have a bottle of glyphosate in my garden shed. I bought it perhaps eight years ago, but have not touched it for the last four, since the IARC study was published. It was the last bottle of pesticide I will ever buy. The more I have learned about them, the more sceptical as to their safety I have become. Regarding glyphosate, I'm not absolutely certain where the truth lies, but it is a no-brainer to conclude that the safest course of action is to leave the stuff well alone.

It seems likely that Europe and perhaps the USA may eventually ban glyphosate over safety concerns, but you can be sure that it will continue to be used with abandon elsewhere in the world. The usual pattern is that new pesticides are first introduced to the market, at a premium price, in the developed world. If they eventually prove to be harmful to the environment or to people they may be banned, but then sales shift overseas, to poorer countries with weaker regulations. For example, paraquat is an old generation of herbicide that was invented by the British company Imperial Chemical Industries (ICI) at its laboratory at Jealott's Hill in Berkshire, west of London. It was first sold commercially in the 1960s, and became the most popular herbicide in the world before the advent of glyphosate. Although intended to kill weeds, it is exceedingly toxic to people. It is a popular choice for suicide, since drinking just a few drops can be lethal, but being so toxic it is also frequently involved in accidental poisonings, particularly in developing countries where pesticide bottles are sometimes recycled as water containers, and farmers may be poorly educated and are often using old and leaky backpack sprayers. Aside from causing death, there is a large body of evidence linking chronic exposure to paraquat to neurological diseases. For example, a meta-analysis of 104 studies found that paraquat use was associated with a doubling of the probability of farmers developing Parkinson's disease. As a result of these concerns, paraquat was banned in the EU in 2007. Even China, a country not famed for its strict environmental regulations, announced in 2012 that it was phasing out paraquat 'to safeguard people's lives'. There can be little doubt that paraquat is a threat to human health. Nevertheless, it is still manufactured in huge quantities, in Huddersfield, in the north of England. It is made in a former ICI plant, now owned by the Swiss chemical giant Syngenta (Switzerland was well ahead of the EU in banning paraquat in 1989). According to the UK's Health and Safety Executive, since 2015 this factory has exported 122,831 tonnes of paraquat, with their records showing sales to Brazil, Colombia, Ecuador, Guatemala, India, Indonesia, Japan, Mexico, Panama, Singapore, South Africa, Taiwan, Uruguay and Venezuela. With striking hypocrisy, EU and UK authorities have decreed that paraquat is too dangerous to use here, but we are quite happy to make and sell it all over the world. According to Baskut Tuncak, a United Nations official who specialises in assessing the safety of hazardous substances, 'This is one of the quintessential examples of double standards.'

As nations, surely we are responsible not only for what happens in our countries, but also for what we export?

The Large Blue Butterfly

Many butterflies within the 'blue' family, properly known as the Lycaenidae, have symbiotic relationships with ants. The caterpillars have glands on their backs that excrete a sugar- or protein-rich liquid much loved by ants. The ants milk the caterpillars regularly, and in return they guard them against predators and parasites.

The largest of the blue butterflies found in the UK, un-imaginatively named the large blue, went extinct in 1979 due to loss of habitat, but was successfully reintroduced in 1984 from Sweden and survives to this day. Fascinatingly, the large blue has subverted its once mutualistic relationship with its host. The female butterfly lays eggs on wild thyme and, after hatching, for the first few days of life the offspring feed upon the leaves like normal caterpillars. Then they do something unusual, dropping from the plant to the ground, and there waiting patiently for a passing ant. They are hoping to encounter a red ant of the species *Myrmica sabuleti*. The caterpillar mimics the scent of an ant grub of this species, so that if a worker ant finds it, she will pick it up and carry it back to her nest, carefully placing the caterpillar in a brood chamber along with the other ant grubs. The ungrateful caterpillar then proceeds to munch her way through the ant brood, the worker ants entirely unable to detect or prevent what is happening right under their noses, despite the butterfly larvae soon growing much larger than any ant grub. The caterpillar remains within the ant nest until the following spring, when it pupates. When the new butterfly emerges from its chrysalis it must quickly climb from the ant nest before it can inflate its wings and fly off, to repeat the cycle.

9

The Green Desert

Plants grow by photosynthesis, the seemingly miraculous process by which they use the energy in sunlight to convert carbon dioxide and water into sugar. To grow, they also need a variety of minerals, which they extract mainly from the soil via their roots. In particular, they need an adequate supply of three elements, phosphorus, potassium and nitrogen, plus traces of many others. Without these nutrients, plant growth is stunted and crop yields are poor. All must be in chemical forms that the plants can access; for example, air is comprised primarily of gaseous nitrogen, but this is no use to most plants as they cannot access it.

Farmers have long understood the importance of soil fertility: there is scientific evidence that Neolithic farmers in Greece were strategically applying manure to wheat and pulses, their most nutrient-hungry crops, nearly 8,000 years ago. Ash and cinders have also been sprinkled on to fields for many thousands of years. In South America, the local peoples have been rowing out to offshore islands to collect guano, the accumulated excrement of sea birds such as cormorants, boobies and pelicans, to use as fertiliser on their land for at least 1,500 years. Guano is unusually rich in phosphates, and was so prized by Inca kings that they introduced the death penalty for anyone who disturbed the birds without permission. The German naturalist and explorer, Alexander von Humboldt, came across the use of guano in 1802, and the subsequent recognition of its value by the western world in the nineteenth century led to the development of a global trade. Thousands of Chinese labourers were shipped to Peru and Chile to mine beds of excrement up to 50 metres deep, surely one of the less pleasant occupations ever devised by man. Of course, such mass extraction could not go on for ever; by the end of the nineteenth century the guano was more or less used up. More recently, the Peruvian government tried to develop sustainable guano harvesting of the fresh excrement, an initiative which tragically collapsed when overfishing removed the birds' food supply, causing their population to crash.

Importing guano from the other side of the world was inevitably expensive, and so scientists in Europe explored other, sometimes bizarre, sources in the desperate search for phosphates. In the 1840s, for instance, the Reverend John Stevens Henslow discovered abundant coprolites, the fossilised faeces of dinosaurs, near Felixstowe in Sussex. Henslow was professor of botany at Cambridge University, and is better known now as the tutor and mentor to the young Charles Darwin. Coprolites had only been recognised for what they were a little over a decade earlier, in 1829, by William Buckland, and Henslow was perspicacious enough to guess that, if bird faeces were rich in nutrients, perhaps fossilised dinosaur poop might also do the job. He developed a technique for extracting the phosphates by treating the coprolites with sulphuric acid, and in doing so he started the little-known coprolite mining rush of the 1860s, Cambridgeshire's answer to the more glamorous Californian gold rush that was going on at the same time. It lasted for thirty years or so but, just as with the guano, it was inevitable that the supply of coprolites would eventually run out; as resources go, they don't get much less renewable than fossilised dinosaur faeces.

As the supply of guano and coprolites dwindled, British farmers then turned to the most unlikely of sources for phosphates: powdered cat. In ancient Egypt, cats were reared

specifically to be slaughtered for mummification. These were
not family pets, sent to keep their beloved owner company in the
afterlife, but mass-reared animals that were strangled or clubbed
to death at about six months old, tightly bound in cloth, dried,
and sold to those who wished to curry favour with one of the
gods. It is suspected that the cats were sold mostly to pilgrims,
who would buy a corpse and then deposit it at the appropriate
temple or shrine, a bit like placing a candle. Thousands of years
later, in 1888, an Egyptian farmer digging near Beni Hassan,
about a hundred miles from Cairo, was surprised when the
ground gave way beneath his feet and he fell into a tunnel
packed with hundreds of thousands of these mummified cats.
Presumably when the temples became over-full with dried cat
remains the priests or, more likely, their slaves, disposed of
them by burying them in the ground. When local farmers began
using crushed-up cat as fertiliser, some entrepreneur hit upon
the idea of exporting this uniquely Egyptian product. Boatloads
of mummified cats were shipped to Liverpool and sold by the
ton at auction, to be ground up into powder for fertilising our
fields.* On one occasion the auctioneer is said to have used one
of the cat's skulls as his hammer.

Meanwhile, in parallel with Henslow, the English entrepreneur and scientist Sir John Bennet Lawes was experimenting
with making fertilisers by treating phosphate-rich rocks, coprolites and animal bones with sulphuric acid. Having patented
his 'superphosphate' fertiliser in 1842, he then spent the following fifty years experimenting with the effects of different
fertilisers on crop growth on his family estate, Rothamsted,

---

* This unusual trade later led to considerable but misplaced excitement
among archaeologists who discovered ancient Egyptian arrowheads in a
Scottish field. There was speculation that the Egyptian empire had perhaps
spread much further than previously thought, or at least that military expeditions from Egypt may have penetrated so far north, until records came
to light showing that the farmer who had owned the field in the 1880s had
purchased crushed Egyptian cats for fertiliser. Quite why arrow heads had
become mixed up with the cats has never been fully explained.

near Harpenden to the north of London*. To this day, phosphate fertiliser is extracted from phosphate-rich rocks, a non-renewable resource that will eventually run out. There are no other sources of phosphates that we might turn to, unless someone discovers a lot more mummified cats. Some even suggest that 'peak phosphate' – the date at which phosphate production begins to decline because of dwindling resources – may be reached as soon as 2030. This is much debated, and others estimate that there are probably far greater reserves of phosphate-rich minerals, though the bulk of world reserves are all in the disputed territory of Western Sahara, currently extracted and exported by Morocco, potentially giving one country a stranglehold on world food production.

The second of the three main nutrients required by plants is potassium. Potassium-rich fertilisers are often called potash, and for thousands of years the main source available to most farmers was wood ash. The clearance of the vast forests of eastern North America for farmland during the nineteenth century provided a huge but temporary supply of potash as the felled trees were burned. Mining of potash-rich minerals from underground began in Ethiopia as long ago as the fourteenth century and, as with phosphates, remains the main source of potash today. Luckily, potash ore is more abundant and more equitably spread around the globe than phosphates, so is unlikely to run out soon.

The third main nutrient for plants is nitrogen, which they require in the form of nitrates, and this occurs only very rarely as a mineable mineral. For millennia, farmers therefore used animal and human manure plus composted crop waste to provide nitrates for their crops, although of course they had no idea what it was about these products that made their crops

---

* Lawes turned his family estate into an experimental farming research station, and today Rothamsted is the oldest such station in existence. One of Lawes' experiments, which he set up in 1856 to study the effects of fertilisers on hay production, has run continuously ever since and is now one of the longest running science experiments ever conducted.

grow so well. Nitrogen was discovered in 1772 by the Scottish physician Daniel Rutherford, but it was nearly a hundred years before the French chemist Jean-Baptiste Boussingault realised the importance of nitrogen-containing compounds for plant growth. Then in 1909, the German chemists Carl Bosch and Fritz Haber developed the Haber process, enabling the capture of atmospheric nitrogen and its conversion into ammonia, which could then be used to manufacture a range of nitrogen-rich compounds utilisable by plants. This was unfortunate timing, as this new process also enabled the cheap mass production of a range of explosives, such as nitroglycerin, nitrocellulose and trinitrocellulose (TNT), just in time for them to be used to obliterate the lives of hundreds of thousands of young men in the First World War. After the war, the huge munitions industry that had sprung up quickly switched to producing fertilisers. Industrial farming was built on the back of poison gas and bombs.

As with pesticides, the use of fertilisers has increased steadily year on year. In the last fifty years the weight of artificial fertilisers applied globally has increased twenty-fold, so that we are now applying approximately 110 million tons of nitrogen each year, plus a further 90 million tons of potash and 40 million tons of phosphate fertilisers.

But why, you may ask, should the manufacture and use of fertilisers be a bad thing? It certainly isn't from the perspective of a hungry human. After all, fertilisers help crops to grow, which seems benign enough. Their increasingly ready availability was certainly a major factor contributing to the 'green revolution', the rapid rise of crop yields around the world in the mid-twentieth century, when new, high-yielding crop varieties were developed that thrived in fertile soils. A particular champion of this movement was the American agronomist Norman Borlaug, sometimes considered the 'father of the green revolution', who promoted the introduction of modern, industrial farming methods around the world, arguing that 'You can't build peace on empty stomachs.' In 1970 he received the

Nobel Peace Prize, and he is credited by some with preventing the death by starvation of a billion people.

As with most new technologies, however, our enthusiasm for the benefits blinded us for some time to the downsides. From an insect's point of view, applications of fertiliser can have devastating consequences. For example, in pasture the application of fertiliser leads to the rapid growth of grasses that outcompete flowers. An ancient flower-rich meadow can therefore be readily destroyed without ploughing or spraying herbicide: a single application of artificial fertiliser will do the job. Much of the south-west of England is bright green when seen from a passing train or from the air, and commuters might assume this 'green and pleasant land', to use the words of William Blake's hymn, is teeming with wildlife. They would be wrong. Much of it is a green desert, a flowerless monoculture of fast-growing rye grass. That is great if you want to produce lots of (monotonous) food for cows, but rubbish if you are a bee or butterfly. The effects are also pernicious, for the leaching of fertiliser into the field margins and hedge bottoms then leads to the dominance of the hedgerow vegetation by a small number of nutrient-loving plants like hogweed, nettles, cock's foot grass and docks. These tall, fast-growing plants squeeze out hedgerow flowers, such as the cowslips that were once so abundant that the flowers could be collected by the bucketful to make wine.

Reduced botanical diversity has inevitable knock-on effects for insects that eat plants and for pollinators. The roadside hedge-banks of the south-west of England are famed for their wildflowers, for example, which might create the impression that the hedge-filled landscape is full of flowers. Yet scientists from the University of Plymouth recently discovered that the sides of the hedges that face farmed fields (which is most of them) have many fewer flowers, and attract far fewer bees than the sides that face the roads.

Forty years ago the wall butterfly (*Lasiommata megera*) was considered an everyday, slightly drab butterfly – mottled brown and orange, with camouflaged grey undersides to its wings – which turned up in almost any sunny habitat in England. When I was a teenager wall butterflies occasionally turned up in our garden in Shropshire, and would live up to their name by basking on the wall of the house. Since then this species has declined by 85 per cent in the UK, and nearly 99 per cent in the Netherlands. A huge hole has opened up in the distribution of this species in the UK, so that it is now absent from much of the Midlands, eastern and south-eastern counties. Patterns of decline seem to correlate with geographic patterns of high fertiliser use (and also with pesticide use), and there is evidence that the lush vegetation resulting from high soil fertility shades and cools the sunny, warm microhabitats that their caterpillars prefer.

Fertilisers may have other insidious and subtle effects on caterpillars. It was recently discovered that caterpillars of several common butterflies, including the small copper, small heath and speckled wood butterflies, are much more likely to die if fed food plants reared in soil with elevated nitrogen. We are not sure why, but a clue comes from a study of tobacco plants, which found that they produce more defensive chemicals such as nicotine when exposed to excess nitrogen dioxide (another accessible source of nitrogen), leading to reduced feeding by caterpillars of the tobacco hornworm (a hawk moth). If food plants become more toxic in conditions of high fertility, then this could easily explain declines of wall butterflies, and is very likely to be contributing to the overall declines of farmland butterflies and moths, and perhaps of other herbivorous insects.

And what about our freshwater habitats, such as rivers, lakes and ponds? The answer is that freshwater habitats draining from agricultural land are similarly polluted with fertilisers (and often also with insecticides and metaldehyde from slug pellets). The excess nutrients allow algae – microscopic plant life – to proliferate, turning clear streams and lakes into a turbid, green soup. This blocks the light from reaching aquatic plants, which die and rot, adding to the nutrients in the water. The rotting vegetation uses up oxygen in the water, which may suffocate animal life. Sometimes, in warm weather, the organisms that proliferate in the polluted water are 'blue-green algae' – technically cyanobacteria, not algae – which release toxins into the water, killing most animal life and even making lakes potentially lethal to human swimmers. All this can be highly detrimental to aquatic life, which, in a healthy river or lake, includes a diversity of insects such as caddisflies, mayflies, stoneflies and dragonflies. Indeed, the relationship between insect diversity and fertiliser pollution in streams is so tight that insects are often used as bioindicators of aquatic pollution. Predictions suggest that the increased frequency of heavy rainfall events under climate change is likely to increase run-off of nitrogen fertiliser and other agrochemicals into rivers, lakes and the sea.

Unsurprisingly, perhaps, our drinking water also frequently now contains fertilisers, particularly in rural areas and developing countries. Nitrates in water are the most common cause of 'blue baby syndrome', a potentially lethal condition in which the pollutant binds to haemoglobin, preventing it from carrying oxygen around the body.

Additionally, the negative consequences associated with raised plant and soil fertility can be found far beyond the farmed environment and the streams draining from it. All burning of fossil fuels in our cars and aeroplanes, or in power plants, also creates nitrogen oxides (NO and NO_2) that rain down on to the soil, increasing nitrogen availability for plants in, for example, nature reserves and other 'protected' areas that may be hundreds of kilometres from any farmland. The manufacture of nitrate fertilisers itself requires a lot of energy, usually in the form of fossil fuels such as natural gas, creating considerable emissions of carbon dioxide. The fertiliser factories have also recently been found to leak large amounts of methane, a greenhouse gas that is thirty-four times more potent than carbon dioxide. Worse still, as much as 50 per cent of the nitrates applied to fields never reach the crop at all, but instead are broken down to nitrous oxide (N_2O)

The Pine Processionary Moth

In the pine forests of southern Europe one can often spot spherical, basketball-sized nests of silk spun high in the branches.

Closer inspection will reveal abundant caterpillar droppings and dried, moulted skins stuck in the silk, and at the heart of the mass a cluster of hairy caterpillars that twitch their tails when disturbed. It is best to investigate with care, as the hairs of these caterpillars can cause a nasty rash. These are the young of the pine processionary moth, so named because the caterpillars set out at night in single file, nose to tail, to walk to a fresh part of the tree to feed, returning to their nest before dawn.

In the early twentieth century the French entomologist Jean-Henri Fabre conducted a famous experiment on the caterpillars, placing them on the rim of a flowerpot, around which they blindly circulated for seven days, each following the other so that they endlessly walked in circles. This study is often held up as an example of the foolishness of blindly following a leader, but it turns out that this is being a bit harsh on the moth. Attempts to replicate it suggest that the caterpillars only stay on the rim of the pot if it has slippery sides that they cannot grip. If they are induced to walk in a circle by containing them within a glass cylinder on a flat surface, almost as soon as the cylinder is removed the caterpillars sensibly break out of the circle and head off to pastures new.

Pandora's Box

Insects naturally suffer the depredations of a broad range of parasites and diseases. Other insects, mites, nematode worms, fungi, protozoans, bacteria and viruses may attack them at any stage in their life cycle. Many of these organisms may sound unappealing – the baculoviruses that cause caterpillars to dissolve from the inside out, for instance, or the mites that live inside the trachea (breathing tubes) of bees – but they are all part of the richness of life. They have coevolved with insects over millennia, and generally did not wipe out their host species (this is rarely a good move for a parasite, since if it loses its host the parasite's demise will inevitably follow close behind).

The most studied insect parasites and diseases are those that infect the domestic honeybee, for the obvious reason that beekeepers try hard to keep their bees healthy and productive and pay close attention to anything that affects their well-being. Perhaps because they are unusual insects, living crammed together in a hive that may contain as many as 80,000 worker bees along with their mother queen – conditions that might seem perfect for the spread of pathogens – honeybees seem to have more than their fair share of diseases and parasites. They are attacked by several species of mite, small hive beetle, fungi such as chalkbrood and stonebrood, bacterial diseases such as American and European foulbrood, wax moths, trypanosomes (relatives of the organism that causes sleeping sickness), microsporidia (fungi-like single-celled creatures), and at least twenty-four different viruses. There are almost certainly more waiting to be discovered. It's a wonder that there are any honeybees left.

All of this is natural, however, and should be no particular cause for concern. These parasites are part of the natural checks and balances that stop any one species from becoming too common. Unfortunately, we humans have inadvertently disrupted these natural relationships by accidentally moving insect parasites and diseases around the globe. We have been beekeeping for thousands of years; as well as being fond of mummifying cats, we know that the ancient Egyptians kept honeybees as they frequently carved images of hives and bees in their hieroglyphs dating back 4,500 years. There is also evidence from North Africa that the practice of beekeeping in clay pots is much older, perhaps dating back as far as 9,000 years. We will never know when humans started moving honeybee hives over substantial distances but, given that they were highly prized, it seems likely that they may have been transported and traded within Africa and Europe for millennia. Indeed, because of this it is hard to know for certain the natural range of the domesticated honeybee (*Apis mellifera)*, but molecular studies suggest an origin in tropical East Africa, with subsequent spread through Africa, Europe and the Middle East. Somewhat confusing, then, that they are often referred to as the 'European honeybee'.

Thanks to human intervention, however, European honeybees are today found in every country in the world, being absent only from Antarctica. They must be among the most widespread organisms on the planet, and their more recent global travels are well documented. The first honeybee colonies were taken to the east coast of North America, for instance, in 1622, and to California in the 1850s. They were also shipped to Australia in 1826, and to New Zealand in 1839. To add to the fun, a particularly aggressive race of the honeybee from Africa was accidentally released in Brazil in 1957, and these so called 'killer bees' have since spread throughout South and Central America into southern USA.

All these transportations were well intentioned. Honeybees give us delicious honey, for millennia the only concentrated source of sugar available to us, and it was for this that Europeans took them around the world. More recently, many more deliberate introductions of other bee species have occurred, motivated by a desire to boost crop pollination. Most developed countries now have more or less strict rules preventing the introduction of exotic species, after disasters in the past such as the plagues of cane toads and rabbits in Australia. However, because bees are seen as beneficial creatures, we foolishly seem to have turned a blind eye to their redistribution around the globe. The United States in particular seems to have an endless appetite for importing foreign bees, having deliberately introduced European leafcutter bees (*Megachile rotundata* and *Megachile apicalis*), and various mason bees (including *Osmia cornuta* from Spain, *Osmia cornifrons* from Japan, and *Osmia coerulescens* from Europe). Why these particular bees were chosen is unclear; it seems to have been opportunistic rather than part of a carefully thought-through plan. All these species and more are now thriving in various parts of North America, to the extent that some scientists now warn American gardeners against putting up solitary bee hotels – artificial breeding homes often installed in gardens to boost native species – because they often become filled with these non-natives. Other bees such as *Chalicodoma nigripes* from Egypt and *Pithitis smaragulda* from India were introduced to the US but seem, so far as we know, to have died out.

Bumblebees have also been spread far from their native ranges. We started with British bumblebees that were shipped to New Zealand as long ago as 1885; four species survive there to this day. One of these, the ruderal bumblebee, was subsequently shipped to Chile in 1982. Bumblebee movements really cranked up in the late 1980s following the development of commercial rearing of buff-tailed bumblebees for tomato pollination. Tomato flowers require 'buzz pollination': the vibration of the male parts of the flower to release the pollen. Most commercial growers rear tomatoes in glasshouses, and up until the 1980s they employed teams of people armed with vibrating 'wands' to buzz each flower. Then, in 1985, a Belgian veterinarian and bumblebee enthusiast named Dr Roland De Jonghe found that placing a bumblebee nest in a glasshouse full of tomatoes provided a remarkable effective pollination service. Bumblebees are adept at buzz pollination, seizing the anthers – the male parts of the flower – in their mandibles, and then buzzing their flight muscles to vibrate themselves and the flower. As buff-tailed bumblebees are also the easiest European species to rear, De Jonghe started raising and selling the nests, and soon set up a company, Biobest. It became a huge industry, with other companies setting up rival factories, and today millions of bumblebee nests are reared annually. Initially the nests were sold within Europe, but the trade quickly became global. Buff-tailed bumblebees turned up in the wild in Tasmania in 1992, escaped from glasshouses in Japan in the 1990s, and were deliberately released in Chile in 1998, from where they rapidly spread across South America. Meanwhile in North America, factories began rearing the common eastern bumblebee (*Bombus impatiens*), a species native to Eastern North America, and shipping them all over the continent and south to Mexico.

With hindsight, all of this movement of bees seems rather foolhardy. Everywhere in the world where one might conceivably grow crops there are numerous native species that would almost certainly be able to do the job, given a little encouragement by the provision of some suitable habitat, and restraint in use of pesticides. For example, North America has about 4,000 species of native bee, plus countless other hoverflies, butterflies, moths, beetles, wasps and more, all busy pollinating flowers. However, there do exist rare instances when a native pollinator was not available for a particular crop. The introduction of long-tongued bumblebee species to New Zealand in 1895 to pollinate red clover,* for example, was necessary as no native pollinators could do the job: clover flowers hide their nectar at the end of a deep tube, beyond the reach of any native bee. However, most of the time these introductions seem to have been entirely unnecessary. They also have the potential to cause great harm.

So what problems might non-native bees cause? The first concern to be raised was that these introduced species may compete with native pollinators, occupying their nest sites (as in the case of the American bee hotels) or taking much of the nectar and pollen so that native species go hungry. As long ago as 1859, Charles Darwin stated that the introduced honeybees in Australia were 'rapidly exterminating the small, stingless native bee'. Darwin was an astonishingly astute biologist, but on this occasion he was wrong: the bee he refers to was almost certainly *Trigona carbonaria*, and it is still fairly common. Nonetheless, Darwin was perhaps the first to recognise the potential for harm, and it seems a good guess that other Australian fauna were impacted by the arrival of honeybees. Australia has about 1,500 species of native bee today, but almost anywhere you care to go, by far the most common insects on flowers are honeybees. There are over half a million hives, containing maybe 25 billion bees – an awful lot of mouths to feed. Australian beekeepers harvest about 30,000 tons of honey per year – all of that honey which, prior to 1826, would have been food for native insects; imagine how many that could have fed.

---

* Red clover itself was introduced to New Zealand from Europe as a fodder crop for livestock, and as a valuable nitrogen-fixing plant that boosts soil fertility, particularly important in the days before cheap artificial fertilisers became available. The British farmers who settled in New Zealand did not initially understand why their clover was setting no seed, forcing them to repeatedly import fresh seed from Europe at considerable expense. It was a solicitor named R. W. Fereday, recently emigrated from England, who worked out the cause of the problem when visiting his brother's farm in the 1870s, a realisation that eventually led to the importation of a selection of bumblebees from England. The full story can be found in my book *A Sting in the Tale*.

Research from around the world has confirmed that honeybees do indeed often impact on native pollinators, displacing them from their preferred flowers, and causing bumblebee colonies to grow more slowly and produce smaller bees in locations where there are many honeybee hives. We will never know the full impact that the introduction of honeybees to places such as North America and Australia has had, since no one was studying native pollinators when the foreign bees arrived. But it remains a possibility that there were once many more pollinator species that we will never know existed.

There is a second, bigger problem associated with the global redistribution of bees by humans: the transported bees often had stowaways. Like Pandora's box in Greek mythology, honeybee hives contained parasites and diseases which have now been spread around the world. Once escaped, they cannot be recaptured. One cannot really blame the early beekeepers for spreading these diseases as, for example, when honeybees were taken to America in 1622 we had not even discovered the existence of bacteria, let alone viruses. Sadly, the inadvertent transport of bee parasites continues to the present day, even though we are now aware of the risks. The global spread of the *Varroa destructor* mite is perhaps the best-known example.

Varroa mites naturally occur on the Asian honeybee, *Apis cerana*, a smaller relative of the European honeybee naturally found, as you might guess, in Asia. The mites are rusty-red, disc-shaped creatures, easily visible to the naked eye at about 2mm diameter, that clamp themselves onto bees and suck out their fat deposits. Although harmful to its host, the mite does not usually cause severe damage to Asian honeybee colonies, the two species having coevolved for millions of years. The Asian honeybee has been domesticated, but has colonies that are generally smaller than those of the European honeybee, and produce less honey. As a consequence, European honeybees were imported to Asia, and are now kept by beekeepers in huge numbers, often alongside Asian honeybees. Unfortunately the *Varroa* mite jumped hosts, developing a taste for the foreign bees: *Varroa* was first recorded on European honeybees in Singapore and Hong Kong in 1963. Careless movement of infested bees and hives around the world then led to the mite spreading westwards to Eastern Europe by the late 1960s, France in 1982, the UK in 1992, and Ireland in 1998. Elsewhere in the world they popped up in Brazil in the 1970s, and the first *Varroa* mite was spotted in North America in Maryland in 1979. Despite rigorous import regulations these relentless little beasts somehow made it to New Zealand in 2000, and to Hawaii in 2007. To date, Australia is the only sizeable country in the world that does not have *Varroa*.

Having not encountered it in their evolutionary past, European honeybees have little defence against the mite. The mites feed on both bee larvae and adult bees, spreading diseases from bee to bee, particularly viruses such as deformed wing virus. As the mites and viruses multiply within a hive, the bees become weakened, and usually, within a year or two, the colony will collapse and die. The *Varroa* mite and its associated viruses are certainly among the bigger problems faced by honeybees, and one of the reasons why beekeepers around the world keep losing their hives.

Luckily, although *Varroa* mites have occasionally been spotted clinging to other insects such as bumblebees, they do not seem able to reproduce on them. It's a shame this is not the case for most other parasites and diseases. For example, deformed wing virus has been found to infect and replicate in insects as diverse as cockroaches, earwigs, social wasps and bumblebees, as well as inside the *Varroa* mite. As the name suggests, infected bees often have crippled, twisted wings, rendering them incapable of flight. Although this virus was first discovered in honeybees and so was long thought to be a 'honeybee virus', there is no particular reason to suppose that it is especially associated with honeybees any more than it is with other hosts – it appears to be a generalist insect virus.

Deformed wing virus was first identified in 1980, in Japan, and has since been detected everywhere in the world anyone has looked, but whether it naturally had a global distribution, or whether it was once a localised insect disease that has since been spread around the world by us, we do not know. What is clear is that honeybee hives now act as a reservoir for this disease, and that it spreads from honeybee hives into wild pollinators such as bumblebees. What level of harm it does is unclear. There is no monitoring of the diseases of wild pollinators, so if an outbreak of disease were to sweep through one, or many, species we would probably not notice. In honeybees, the virus can persist in colonies without doing much harm, but in combination with *Varroa* it seems to become symptomatic, with heavily infected bees having deformed and useless appendages, most crucially their wings. It is not uncommon to find nests of bumblebees in which some of the bees have deformed wings, and some bumblebees with deformed wings have been tested and proved to be positive for the virus, but how many wild bumblebees are harmed in this way we have no idea. And this is just one such disease, the best known of the viruses found in honeybees.

At least one other bee disease has certainly spread from honeybees into wild pollinators. *Nosema ceranae* is a microsporidian, a single-celled organism that infects the guts of bees. Like *Varroa*, it seems that it was originally a parasite associated with the Asian honeybee, and it was first discovered as a new species as recently as 2004, in Taiwan. As soon as researchers started looking for it, they found *N. ceranae* to be very common in honeybees across Europe and North America. It is widely regarded as an emerging disease in these regions – i.e. one which has recently arrived – but in truth it is very hard to be sure where it came from. Genetic studies of old specimens of honeybees have detected *N. ceranae* in the USA in 1975 and Brazil in 1979, so as with deformed wing virus it is unclear where its origin lies, and when it spread around the world; scientists are still trying to piece together what happened, but it is certainly too late to reverse the damage.

Also in common with deformed wing virus, effects on the host seem to be variable, perhaps depending on how healthy the bees are in other respects. Some studies have found that *N. ceranae* kills honeybees within eight days, with foraging bees being most strongly affected, so the colony often ends up with a queen and her nurse bees (the younger workers that look after the brood) but few experienced foragers to bring back food. As with deformed wing virus, *Nosema ceranae* also seems to turn up frequently in bumblebees, and has now been found in wild bumblebees in China, South America and the UK. It seems to have higher virulence in bumblebees than in honeybees, often causing death, so recent evidence that about one quarter of free-flying wild bumblebees in the UK are infected with this disease is pretty worrying.

The spread of *Varroa*, and probably also of *N. ceranae,* is almost certainly due to the careless movement of honeybees, but the trade in factory-reared bumblebee colonies has further added to the mess. Hygiene in the bumblebee factories seems to have improved in recent years, but when the industry started in the 1980s and 1990s many of the nests that were shipped out contained parasites of one sort or another. Tracheal mites from Europe, hidden within the breathing tubes of the factory bees, have escaped in Japan and now infest native bumblebees there. Similarly, buff-tailed bumblebees introduced to Chile from Europe have been spreading across South America and seem to be carrying at least three bumblebee diseases, although it has not been established beyond doubt where these diseases originated from. What is certain is that the arrival of the buff-tails seems to have had a devastating impact on the native Andean 'giant golden bumblebee', *Bombus dahlbomii*, thought to be because the native bee has no resistance to the parasites (see my book *Bee Quest* for further details).

We are continually discovering new bee diseases, although we have only a rudimentary understanding about their natural host range and geographic range and the basics of their biology. However, we know far more about them than we do about the diseases of other insects. Bees are not the only insects we have moved about the globe. For example, the companies that rear bumblebees also rear a wide range of biocontrol agents, including tiny parasitoid wasps, predatory mites, hoverflies, lacewings and ladybirds, all of which are shipped around the world without much regard for the consequences. We do not know what diseases they may carry. We also accidentally move insects on a vast scale with ornamental plants, which are still routinely shipped across international borders. Other insects stow away in shipping containers: for example, Asian hornets arrived in Bordeaux in a shipment of ornamental pottery from China, and soon spread out to colonise most of Western Europe. It seems certain that our activities, particularly over the last couple of centuries of extensive global trade, have profoundly altered the distributions of numerous insect parasites, but we will probably never know what impact it has had. Our ignorance of insect diseases, and what impact they have on their hosts, is impossible to exaggerate. For 99.9 per cent of insect species we know simply nothing. It is entirely plausible that epidemics of non-native parasites or diseases of insects, accidentally spread by man, have ravaged populations of insects, and may be continuing to do so, without us noticing anything at all.

Suicide Bomber Termites

In the remote, steamy forests of French Guiana lives a most unusual insect, which goes by the name of *Neocapritermes taracue*.

Termites are all fascinating creatures, unrelated to ants but living a similar lifestyle, in sometimes vast underground colonies with a queen, and worker castes specialising in different jobs. Unlike ants, however, termites are strictly vegetarian, feeding on all manner of plant materials and relying on symbiotic microbes in their guts to break down otherwise indigestible cellulose. Termites are the favoured food of many creatures, from giant anteaters to ants themselves, and they have evolved many different defences over the millennia, but none more committed than that of *Neocapritermes taracue*. As they become old, workers of this species develop blue spots in their abdomens, filled with copper-rich proteins. Over time their jaws also wear out and become blunt, and they are limited use to the colony, but these workers become more aggressive, and attack any intruder. If the fight isn't going well, the old workers burst, and the blue proteins react with hydroquinones stored in the salivary glands of the now-dead insect to form highly toxic benzoquinones (the same compounds that blast out of the bottoms of bombardier beetles). Scientists describe this behaviour as suicidal altruism, analogous to the barbed sting of worker honeybees which results in the bee's death if it stings. In both cases, the worker is willing to nobly sacrifice itself for the greater good of its colony.

The Coming Storm

Of all the great environmental challenges facing humankind in the twenty-first century, climate change is perhaps the most familiar and undoubtedly one of the most pressing.

Yet it was not until as recently as the 1990s that scientists reached a consensus that the greenhouse gases produced by man's activities were altering our climate. Even today there are deniers, sadly including the previous President of the United States and many of his followers – but then there are also people who argue that the world is flat.* For those of us living in the real world it is clear that we are changing the climate, and that it is happening fast. So far, the climate has warmed by about 1°C since 1900, which doesn't sound like much. It

---

* It is easy to laugh at this, but it has a serious side. There really is a Flat Earth Society based in the UK with offshoots in Canada and Italy. Its members believe that the earth is a disc, with the North Pole forming a hole in the centre and Antarctica actually being a 150-foot-tall rim around the outer edge. They maintain that there is a worldwide conspiracy by governments to convince us that the Earth is spherical. The echo chamber of social media seems to encourage these crazy beliefs, so that membership of this society has grown in recent years to more than 500 – although I'd guess that not all members are entirely serious. Nonetheless it should surely be a cause of concern that significant numbers of people can come to believe something that is so obviously nonsense, and it makes me wonder what a real well-funded conspiracy might be able to achieve (for example in rigging elections, or greenwashing practices that harm the environment).

is roughly the equivalent in temperature change of moving house from Birmingham in the centre of England to Brighton on the South Coast, although because the change has taken place over a hundred years or so no ordinary person living *in situ* would have noticed anything happening; a nice example of a shifting baseline. On the other hand, I was born just north of Birmingham and have moved to near Brighton, and I can notice the difference, because I experienced the change quickly.

Of course, moving from Birmingham to Brighton, in terms of weather, doesn't sound too bad at all to those of us that live in a cool, temperate climate, and I've always suspected that this underlies the apparent apathy about climate change among many people in the UK and in other damp and chilly countries. But of course this is entirely to miss the seriousness of the threat of climate change. Unless we act very soon to massively reduce emissions of greenhouse gases we are likely to see average temperatures rise by 3°C to 4°C within the lifetimes of today's children. Temperature rises will not be

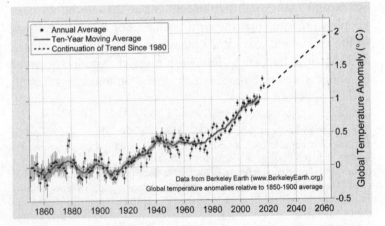

Global temperatures from 1860 to the present, with projections to 2065: At the current rate of progression, the increase in Earth's long-term average temperature will reach 1.5°C above the 1850–1900 average by 2040, and 2°C will be reached around 2065. From http://berkeleyearth.org/global-temperatures-2017/.

distributed equally: most predictions suggest that temperatures over the majority of the oceans will rise little, with temperature increases far higher than the average at the poles, and over large continental land masses, particularly Africa and Asia. Many parts of Africa and Asia will therefore become more or less uninhabitable for humans, and of course for most wildlife too, and these are the very places where the majority of the human population resides.

Attempting to predict the course of climate change is probably humanity's biggest single scientific endeavour of all time, occupying the minds of literally thousands of scientists around the world over decades. The science is not precise, particularly because climate is affected by complex 'feedback loops' that can be hard to understand or capture accurately in mathematical models. Positive feedback loops will accelerate climate change, and some predictions even suggest that global warming could become a runaway process that we simply cannot stop, perhaps as soon as 2030. For example, the ice and snow at the poles and in mountainous regions reflect the sun's heat, helping to reduce warming. As they melt and recede, less heat will be reflected, resulting in faster melting, and so on. That is a positive feedback loop. Similarly, the thawing of the frozen tundra in the far north results in the bubbling up of methane that was created over millennia by the very slow, anaerobic rotting of organic matter, but which until now has been trapped beneath the ice. Methane is a much more powerful greenhouse gas than carbon dioxide, so more methane means more warming means more methane, and so on … Warming of soils speeds up the rate at which organic matter oxidises to carbon dioxide, simultaneously reducing soil health while adding to greenhouse gas emissions. Warming is also likely to make forest wildfires more frequent, which of course very rapidly turn trees into carbon dioxide and smoke.

On the other hand, there are some negative feedback loops that might slow climate change, not least of which is that, as the Earth gets hotter, so more heat is radiated into space. Higher carbon dioxide levels may help plants to grow faster, absorbing more carbon dioxide as they do so. There are many more such loops, both positive and negative, some clearly established and others more speculative, but the strong consensus among climate scientists is that the effects of the positive feedback loops are likely to far outweigh the negatives. Overall, there is no doubt that it is going to get warmer, but exactly how fast is unclear, and of course a lot depends on what we do next.

Even harder to predict are other aspects of our future climate, particularly the strength and frequency of rainfall and of extreme weather events such as hurricanes. Warming means more evaporation of water from the surface of the land and sea, and since what goes up must come down this inevitably means more rainfall, and in particular more heavy downpours, leading to more flooding. We have already seen an increase in the frequency and strength of hurricanes in the Atlantic, some of them having devastating effects on the southern USA and the Caribbean. Most models suggest profound changes in the places where rain falls by the end of the century: although there will be more rain overall, there will be changes in where it falls, and some places are expected to have less rain. The Sahara is predicted to expand northwards into southern Europe and south into much of equatorial Africa, while parts of the Amazon basin are likely to become drier, so that whatever rainforest is left is likely to die anyway.

Projected sea level rises are also the subject of much debate. There is no doubt that the layers of ice that cap land near the poles and the world's great mountain ranges are slowly dripping into the sea. In 1850 Glacier National Park in Montana had 150 glaciers. Today it has just twenty-six. Soon it may need a new name. At the same time, the water already in the sea expands as it warms. Most estimates suggest that sea levels may rise by about 1 to 2 metres by the end of this century. Two metres might not sound much, but it would entirely obliterate the Maldives, the Marshall Islands, most of Bangladesh (home to 168 million people, and one of the most populous countries on Earth), much of Florida, and many huge cities such as Jakarta and Shanghai.

By 2100, in the USA alone, an estimated 2.4 million homes will be under water. However, there are realistic concerns that sea level rises may be much greater. Some studies suggest that we are near a tipping point beyond which the Greenland ice sheet will inevitably melt; that alone would raise sea levels by 6 metres. Most predictions about climate change focus on what the world will look like between now and the end of this century, but sea levels will continue to rise for many centuries beyond, even if the temperature stabilises. This is simply because vast blocks of ice are slow to thaw; parts of Antarctica are buried under ice 4 kilometres deep. If we burn all of our fossil fuel reserves it is estimated that it would be enough to eventually melt the entirety of the Antarctic ice, although it might take 1,000 years for it all to inexorably trickle into the oceans. That would raise sea levels by 58 metres, leaving rather little land for whatever terrestrial life remains.

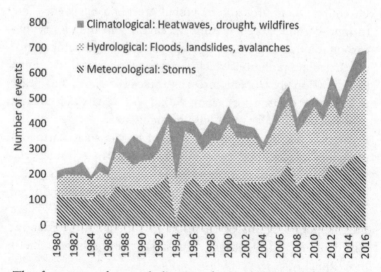

The frequency of natural disasters from 1980 to 2016: Natural disaster loss events due to floods, storms and fires have more than tripled in frequency since 1980. Data are based on insurance losses. Disasters that impact on humans will also have profound effects on insects. Data source: *Economist*.

One way or another, it seems that flooding events will become steadily more frequent, either due to heavy rainfall causing flash floods as streams and rivers burst their banks, or in coastal areas from sea level rises and increased storm surges.

While floods may threaten many places, others are more likely to burn. In 2019 California and several Mediterranean countries, such as Spain for instance, saw unprecedented numbers of wildfires. If they are dried by artificial drainage or drought, peat soils can ignite too: for example, in 1997–8 unusually low rainfall during an El Niño event resulted in wildfires in the rainforests of Borneo and Sumatra which spread into the peaty soils, burning for months and destroying six million hectares of rainforest and releasing about two gigatons of carbon dioxide.[*] Further peat fires in South East Asia took place in 2002, 2013, 2014 and 2015. In Brazil the recent fires have been mostly deliberate, encouraged by the populist President Bolsonaro, keen to open up the Amazon to farming and mining. Of course, these fires add to global greenhouse gas emission, turning carbon stored in timber, leaves, soil and sloths into carbon dioxide and clouds of polluting smoke (a major source of particulate pollution which is extremely harmful to human health). Globally, tropical forest clearance is causing the release of about 4.8 gigatons of carbon dioxide per year, which equates to 8 per cent of all greenhouse gas emissions.

We may not be surprised to hear of fires in such warm and often dry places, but 2019 also saw vast tracts of Siberia, Greenland and Sweden improbably in flames, with the heat eventually drying and igniting peat beds that then smouldered through the summer, almost impossible to put out. These fires in polar regions often result in soot landing on snow, darkening it and increasing heat absorption so that it melts more quickly.

---

[*] To put this in context, total global carbon dioxide emissions from all burning of fossil fuels in 1997 were about 24 gigatons, so these peat fires alone added about another 10 per cent.

Some places are likely to experience both fires and floods, though perhaps not at the same time. Heavy rain in 2018, falling on hillsides that had previously been denuded of vegetation by wildfires, led to devastating mudslides in California, for instance, killing twenty-three people as thousands of tons of mud and boulders slid down the steep slopes.

It is clear that climate change is likely to pose massive problems for us humans, but what about the impacts on insects and other wildlife? Until recently, direct evidence that climate change has already had major impacts on insect populations was not strong. The ranges of some insects have begun to shift in response to climate, with European and North American bumblebees tending to disappear from the southern edges of their range, and move to higher elevations in mountainous regions. There is also evidence that the timing of the emergence of some herbivorous and pollinating insects in the spring is becoming decoupled from that of their host plants: for example, some mountain plants in Colorado are now coming into flower before the bumblebees that feed on them have emerged from hibernation, when previously they did not. Perhaps if changes occurred slowly, the bees or the flowers might be able to adapt. So far, the changes they have experienced have been fairly subtle, but the pace is expected to become much faster as climate change accelerates through the twenty-first century.

Whereas most insect groups are at their most diverse in the tropics, bumblebees as a whole tend to be found in relatively cool climates. They are big and furry as an adaptation to keeping warm. Intuitively, it therefore seems likely that climate change may be particularly bad for them. Climatic predictions can be used to map the future possible distributions of wildlife, based on the range of climatic conditions they occupy today. Essentially, computer models can work out exactly what range of climates are currently occupied by any particular species, in terms of annual patterns of temperature, rainfall and so on, and then calculate where in the world is likely to have that climate in the future. As one might guess, ranges of almost all species are predicted to shift away from the Equator. These calculations have been done for all European bumblebees (and for many other creatures), and as you might expect there are predicted winners and losers. *Bombus mesomelas*, for instance, is a pretty bumblebee with an ashy grey thorax and golden bottom, found in flowery mountain meadows in southern Europe. It could, theoretically, move to the UK in the future. On the other hand, familiar species such as the common carder bumblebee, early bumblebee, red-tailed bumblebee and the garden bumblebee are predicted to disappear from lowland Britain and much of Europe by 2080, though they should be able to cling on in Scandinavia and Scotland. The tree bumblebee, which only arrived in the UK in 2001 from continental Europe, is predicted to be gone again by the end of the century.

What about other groups of more typical, thermophilic insects, such as butterflies? Warming might be expected to be a boon for them, at least in temperate countries such as the UK where many species are near the northern edge of their range. Martin Warren and colleagues from the charity Butterfly Conservation analysed changes in the populations of forty-six butterfly species that all reach the northern edge of their range in the UK; the species we would expect to be enjoying warming. Between 1970 and 2000, three-quarters of these species declined significantly. The pattern differs between sedentary habitat specialists (fussy species with very specific requirements and low mobility, comprising twenty-eight species) and the generalist, highly mobile species (eighteen species). Of the habitat specialists, 89 per cent had declined, while only half of the generalists had declined and a few were thriving. This gives us a clue as to why climate warming has so far not benefited even warmth-loving butterflies. Mobile generalists can more easily move in response to warming, and are more likely to find somewhere where they can survive when they get there. By contrast, the habitat specialists tend to be less mobile, and even if they do manage to make the journey they will die anyway if they do not find the right habitat when they arrive.

Here it is worth pulling back briefly to reflect that the climate has always changed, and that shifts in the distributions of species are a natural response that has been occurring for millions of years. Everything, from bumblebees and butterflies to oak trees and reindeer, has moved, more or less easily, as ice ages have come and gone. An oak can drop an acorn a few metres north, or be very lucky and have a jay carry its acorn a few hundred metres, so that over the course of 10,000 years a succession of trees might edge 100 kilometres towards the pole. For bumblebees and butterflies it should be much easier, for they have wings. The trouble is, this time climate change is happening very fast, and it is occurring in a world when natural habitats are already badly degraded and patchily distributed. As a result, most butterflies and bumblebees don't seem to be moving north. While they are disappearing from the southern edges of their range in Europe and North America, the expected advance at the northern edge does not seem to have happened, with the exception of a very small number of species. Additionally, species with low dispersal abilities, such as oak trees, snails or woodlice, need a more-or-less continuous habitat if they are to edge slowly north or south, and in the pre-human world that would have been fine. Nowadays, with much of the land under intensive farmland and most of the rest covered by roads, golf courses, housing estates or factories, it is much harder for wildlife to move. There are fewer places where an acorn might fall where it is ever likely to grow into a full-sized tree and produce acorns of its own. Even flying creatures such as bumblebees, which can readily zoom over a motorway or an arable field if they so choose, need to find somewhere suitable to live when they get to the other side. The reality today is that many creatures are now surviving as much-reduced populations in little pockets of more-or-less isolated habitat: nature reserves and the like. The odds of them successfully leapfrogging northwards to keep ahead of climate change are slim, particularly since they are reliant on the community of flowers they feed upon to move with them.

For instance, the climate models predict that large chunks of Britain have recently become suitable for occupation by the silver-studded blue butterfly, a handsome but weak-flying little purplish-blue butterfly with gorgeously spotted cream, orange and black undersides to its wings. However, the butterfly has not swept majestically northwards, but remains instead hunkered down on a few patches of heathland, its favoured habitat, in the south of England. The climate may be suitable further north, but there are few patches of heathland and little chance of it getting to them under its own steam. Similarly, the aforementioned bumblebee, *Bombus mesomelas,* currently lives in the hilly meadows of Italy and neighbouring countries. In theory the climate of the southern UK might well suit it by the end of the century, but how will it get here? And if it does, will a suite of suitable Italian meadow plants have arrived to welcome it (very unlikely), or will it be able to adapt to whatever plants it finds (possible). It seems more likely to me that both the silver studded blue and *Bombus mesomelas* will simply fizzle out *in situ* as their habitats slowly become too warm for them.

The climatic models that predict possible future distributions of species tend to be based on averages – monthly mean temperature, mean rainfall and so on. What they cannot account for are the effects of extreme weather events, such as droughts, heatwaves, wildfires, storms and floods, all of which are likely to become more frequent and more extreme in the future. We have very little idea what impact these will have on insects, but of course very few of them will be positive. Fires will obviously kill insects, but the flush of new flowers that follows fires in some ecosystems would benefit some. Summer storms are likely to batter delicate adult insects such as butterflies, and flash floods are likely to destroy underground nests of creatures such as bumblebees. Drought causes water-stressed plants to cease nectar production in their flowers, which will certainly harm pollinators, while cold-loving insects such as bumblebees will overheat and be unable to forage in heatwaves. In prolonged droughts plants wilt and become unpalatable for caterpillars;

for example, in the hot British summer of 1976 many caterpillars of the Adonis blue butterfly died as their food plant,
horseshoe vetch, shrivelled in the heat. As a result, numbers
of the adults were much lower the following year, and some
populations died out. Insects have, of course, coped with all
of these events in the past, but the increased frequency and
strength of extreme events at a time when many insects have
already declined may be the final straw for some.

Yet although climate change is broadly bad news for most
creatures, it may benefit a small number of insects. Tough,
mobile, adaptable creatures such as house flies, which breed
in the excrement of humans and of our livestock, and in the
festering used nappies thrown into landfill, will be able to
breed faster in the warmer future. The ever-increasing numbers
of humans and our livestock will mean more food for them.
Warming will allow pest insects to get through more generations per year, which will allow them to evolve resistance
to pesticides even faster than before. They will reach larger
population sizes before being knocked back by winter, and
as winters become milder so some pests will become able to
thrive all year round. The great wheat belt of North America
is currently situated in a more-or-less optimal climate for the
crop (not by coincidence), and even without considering insect
pests it is predicted that crop yields will fall by about 10 per
cent for every degree of warming. On top of that, the accel-
erated multiplication of pests such as aphids and caterpillars
is expected to decrease yields by a further 10 to 25 per cent
for every degree of warming, an estimate that also applies to
other global staples like rice and maize.

In addition to crop pests, any organism that can cope well
with life in urban areas is likely to prosper in the coming years,
for the extent of urban habitats will inevitably increase as the
human population heads towards ten billion or more. The
yellow fever mosquito (*Aedes aegypti*) seems to have adapted
well to urbanisation and thrives in cities, breeding in blocked
gutters, discarded tyres, barrels, buckets and any other human

refuse that traps puddles of water. It is one of the main vectors of several nasty diseases, including dengue fever, chikungunya, Zika fever and of course yellow fever, as its name suggests. The *Anopheles* mosquito, the main transmitter of malaria, is also benefiting from the spread of man's activities. Cases of malaria tend to become more frequent in areas where forests are cleared for agriculture, because the mosquito likes to breed in sunlit puddles and ditches, which are hard for it to find in dense forest. Climate predictions suggest that malaria is likely to spread to higher-altitude regions of the tropics, for example in Colombia, Kenya and Ethiopia. These regions are densely populated with humans in part because, until recently, they were largely free from malaria. The southern states of the USA, south-eastern Europe, parts of China and the densely populated areas surrounding São Paolo and Rio de Janeiro in Brazil are all likely to become suitable for malaria by 2050. Dengue fever is similarly predicted to become far more common throughout North America, as far north as southern Canada. One estimate suggests that the number of people at risk from viral diseases spread by the yellow fever mosquito and the re-lated Asian tiger mosquito (*Aedes albopictus*) will increase by one billion by the end of the century (not accounting for rising human populations). The only good news is that some lowland equatorial regions may become too hot for the transmission of malaria – but they will probably also be too hot for humans to survive in anyway.

It seems certain that climate change will have profound im-pacts on insects in the future, but can it explain their declines to date? The authors of the Krefeld insect study specifically investigated whether changing climate could be the cause of the dramatic 76 per cent decline in insect biomass on German nature reserves. Although day-to-day weather patterns had big impacts on the numbers of insects caught – as you would guess, insect catches are higher on sunny days – the overall climate in Germany did not change much over the relatively short twenty-six-year period of the study. Climate change could not explain the decline, the authors of the study concluded, and there was little disagreement from the scientific community, there being plenty of other potential culprits.

In 2017, the same year that the German study was published, Sarah Loboda of McGill University in Canada published data on changing populations of the fly species found on Greenland, tough insects adapted to the cold, windy conditions and very short summer. The paper received scant attention, perhaps because no one much cares about flies, but it describes an 80 per cent drop in overall abundance over a nineteen-year period to 2014: slightly more rapid than the German paper. Loboda ascribed this population collapse to climate change, which has been more marked at the poles than elsewhere; in Greenland, other man-made impacts are negligible. Then, in 2018, Brad Lister's Puerto Rico study was published, and climate change came back into the frame as a contender to explain insect declines. Lister, you may recall, had sampled insects in a rainforest in 1976 and 1977, and returned to the same sites after thirty-four years to repeat the exact same sampling in 2011–13. He found that the insect biomass had declined by 80 per cent for samples collected via sweep nets, and a staggering 98 per cent for insects caught on sticky traps. These forests have not been logged or otherwise directly altered by humans in the last thirty years, and no pesticides have been used on or near them so far as anyone knows (which is also true of the Greenland study sites). Ostensibly the forest remained entirely unchanged, except that almost all the insects had vanished. Unlike in Germany, however, according to a weather station within the forest the climate has changed since the late 1970s, with an increase in the mean maximum daily temperature of 2°C. Lister tentatively concluded that this was the most likely cause of the Puerto Rican declines. As one of the 'referees' of this paper – meaning I was asked to evaluate the quality of the paper as part of the 'peer-review' process – I could not think of a better explanation, although I was not wholly convinced. Just because the temperature was the only factor that seemed to have changed does not mean that it must be the cause of insect population collapse. There could, for example, have been a devastating epidemic of some unknown insect disease, contamination of the forests with an unidentified pollutant, or a visit to the forest by an army of insect-eating aliens (OK, I'll admit this one is not particularly likely). My point is only that there are many other possible explanations.

The Lister paper attracted a lot of scrutiny, and it emerged that there was a flaw in the evidence that the temperature had actually changed. The recording equipment had been replaced some time between Lister's two visits to Puerto Rico, and the hike in temperatures, rather than being a gradual process, seemed to coincide with the switch in equipment. In other words it seems likely that the climate hadn't actually changed as much as the records showed, but rather that the apparent jump in temperature was at least in part an artefact of the changing methods used to measure it. Some critical reviews were published, suggesting that the paper was fatally flawed, but if we put the climate issue to one side, the core findings of the paper – that there had been a massive decline in insect abundance – remain unchallenged. The explanation, however, is still disputed.

For the moment, the future climate of our planet is still in our hands. We have already profoundly altered it, but if we were to act decisively we could prevent it becoming much worse. The dire projections of what the world might look like by the end of the century need not come true. In 2016, 196 governments from around the world made commitments to keep climate change below a maximum of 2°C, and ideally no more than 1.5°C above pre-industrial levels in the Paris Agreement. In the years since, not one single major industrial nation has got on track to fulfil those commitments. All the measures put in place so far to tackle climate change, such as the proliferation of green energy sources (wind, solar, wave etc.), the move to more fuel-efficient cars, better insulation of homes and so on have had no measurable effect on carbon dioxide emissions, which continue to rise *at an accelerating rate* every year.* We have simply been using more energy than ever before, more than wiping out the benefits of these new technologies. You might expect that creating green energy would reduce the need for energy from fossil fuels, but it has not worked out that way so far. Instead, our energy-hungry economy simply sucks up all the energy it can get, and cries out for more.

In the meantime, Donald Trump pulled the USA out of the Paris Agreement (although fortunately Joe Biden reversed this on his first day in office). You can see for yourself how countries' efforts to combat climate change compare on the Climate Action Tracker website (https://climateactiontracker.org). Only Morocco and the Gambia are on track to meet their promises in the Paris Agreement, two tiny countries with almost zero emissions to start with. Countries whose efforts are woefully inadequate, and likely to see us heading towards global warming of 4°C or more (catastrophic for all life on Earth), include the USA, Saudi Arabia and Russia. It is perhaps not a coincidence that these three countries happen to be the three biggest oil producers in the world. One might be forgiven for suspecting that their heart is not really in tackling climate change at all. In the case of the USA this was abundantly clear under the Trump administration. Interestingly, Russia has other reasons to ignore the perils of climate change, for many Russian ports currently become unusable during the winter months due to sea ice, but will soon be ice-free for much or all of the year. In addition, vast tracts of northern lands that are currently too cold for crop production will become suitable for growing cereals,

---

* At the time of writing, in November 2020, the world is in lockdown due to the coronavirus epidemic. This is likely to put a small, temporary dent in greenhouse gas emissions.

so that Russia may be able to step into the gap created when the USA wheat crops start to fail. We should not expect help in tackling climate change from Vladimir Putin any time soon.

The fundamental problem with the Paris Agreement is that it has no teeth at all. It relies entirely on countries choosing to cut their own emissions, with no penalty if they fail. It is very easy for a government to make a long-term promise, knowing that different politicians will be in charge by the time any reckoning is due. One only has to look at the 1992 Rio Convention on Biological Diversity, which was signed by almost exactly the same group of 196 governments as signed the Paris Agreement. In the Rio Convention our governments promised to halt the loss of global biodiversity by 2020. In reality, the period 1992 to 2020 has seen the greatest loss of global biodiversity for at least 65 million years. We cannot rely on the empty promises of our governments to save our planet.

Vein-cutting Monarchs

The monarch butterfly of North America is famed for its beauty and for its extraordinary migration from Canada to over-wintering sites in Mexico, but its caterpillars have their own remarkable behaviour. They feed upon the leaves of milkweed, a plant named after the white, sticky sap that flows from damaged leaves, also known as latex.

Latex is produced by many plants, and some types are harvested to make into rubber. Naturally, latex has two purposes: it dries and blocks the wound, acting like a scab, and it sticks to and poisons any herbivore attempting to nibble the plant's leaves. It successfully deters many insects, but a few, including the caterpillars of the monarch butterfly, have worked out how to overcome this defence. The caterpillar simply chews a trench across the base of a leaf, severing the latex-containing vessels and allowing the latex to drain out, so that it can then enjoy consuming the remaining, undefended portion of the leaf.

Bauble Earth

We have all seen the dramatic satellite images of the Earth at night, its land masses etched with an orange glow from billions of electrical lights, hanging like a Christmas bauble in space. North America, Europe, India, China and Japan in particular are lit up with a festive light. Every city is visible, with the great metropolises gleaming brightest, and most coastlines glimmering with a ribbon of human development. Very few places on land are truly dark, aside from the frozen wastes near the poles, the great deserts, and a few remaining parts of the Amazon and the Congo. It is estimated that the amount of light we are casting at night increases every year by between 2 and 6 per cent. Every day the human population increases by about 225,000: enough for a new city every night, for a new gleaming spot of light visible from space.

Most of the nature reserves on which insects were trapped in the German study are reasonably close to urban areas. Even if bright lights are not always directly visible from the reserves, skyglow, as we call the scattered light pollution visible in the atmosphere above cities, can be seen hundreds of kilometres from its source. Some scientists have argued that light pollution may have contributed to the dramatic decline in insects. But is this a plausible explanation?

Let's consider what harm light pollution might do to insects. More than 60 per cent of invertebrates are nocturnal, and most use light from the stars or Moon as a cue for navigation and orientation. Artificial lights attract flying insects such as moths and flies. Clearly disorientated, they bash themselves against the light, becoming burned or damaged, exhausting themselves and making themselves vulnerable to predation. I vividly recall camping out many years ago on a campsite in tropical Australia which had low-level lighting on posts to illuminate the paths to the toilet block. At night, every light had at least one fat cane toad squatting beneath it, hoovering up the near-endless supply of insects that blundered into the lamps. In Spain, similarly, I have seen hungry geckos crowd around lights to help themselves to the dazzled insects. Spiders often spin their webs on and under outdoor lights, the webs sometimes becoming thick with flies and other small insects, while bats may take advantage of the clouds of confused insects to swoop in and snatch prey for themselves. In the absence of cane toads, spiders, bats or geckos, come the morning any surviving insects are often left sitting, dazed, in an exposed position on a lamp post or nearby wall where insectivorous birds find them easy pickings. Any keen moth trappers reading this will know that they need to empty their traps at first light or else wrens and tits will quickly learn to climb into the trap and partake of a sumptuous breakfast, leaving nothing but a pile of wings.

Why insects are attracted to lights at night has never been fully explained. Moths, after all, do not naturally try to fly to the Moon. There are competing theories, the most popular and convincing of which is based on the notion that insects navigate using the Moon when migrating. An insect intending to travel a long distance in a straight line might fly at a fixed angle to the Moon, gently adjusting the angle as the night progresses and the moon moves across the night sky by using some sort of internal clock. Bees use the sun in a similar way when navigating to and from patches of flowers. The theory is that nocturnal insects mistake a bright light for the Moon, but because it is close to them rather than thousands of miles away, their angle to the light shifts very rapidly if they fly in a straight line. To compensate, they curve towards the light, flying in a decreasing spiral until they crash into the lamp.* Regardless of the explanation, there is no doubt that our artificial lights lead countless billions of insects to an untimely demise each year. Every light is a population 'sink', steadily sucking insects from the landscape night after night.

It is also likely that our lights lead to more insidious problems. Some insects may not blunder into lights, for instance, but might nonetheless be disorientated. We know, for example, that dung beetles can detect the hazy line of the Milky Way in the sky, using it to help orientate themselves and maintain a straight path when rolling their balls of dung. We do not yet know what confusion it might cause to the beetles if their route passed near an artificial light. Perhaps more importantly, most insects, including those that are primarily active in the daytime, use light as a critical cue in triggering biological clocks. For example, many organisms use day length for timing their life cycle, so they emerge from hibernation, or lay their eggs, at an appropriate time of year. Some insects time their foraging activity to coincide with, or avoid, the time of the month when the Moon is at its brightest. Mayflies time their emergence as adults to coincide with the full Moon. The timing of these cycles is vital: consider what would happen if a mayfly, which lives for just a few hours or days at best, times its emergence incorrectly. Then it will have no-one to mate with, and will die alone, without fulfilling its reproductive potential. There have been few scientific studies of the effects of light pollution on insect cycles of this sort, but it seems highly plausible that insects living near bright lights might mistake a street lamp for the full moon, or the rising sun, and get their timing disastrously wrong.

For a few unusual insects, our lights may pose a particular obstacle when it comes to finding a mate. Fireflies and glow-worms emit light to attract a member of the opposite sex.

---

* You can read more about this and other competing theories in my previous book, *The Garden Jungle*.

In the European glow-worm, for example, the female insect (they are beetles, not worms) emits a charming greenish glow from her bottom which males find irresistible. For millions of years their luminous bottoms must have been easily spotted in the dark of the night, but now they have to compete with the far brighter lights of humankind. For a male glow-worm, finding yourself attracted to the bright lights of the city would be a disaster.

The risk posed by artificial light is also likely to depend on the type of light being used. Moth collectors have long known that lights emitting lots of ultra-violet are most attractive to moths. In Europe, street lights were until recently mainly high-pressure sodium lamps or mercury vapour lamps that emit a lot of UV. To save energy, many are being replaced by light-emitting diodes (LEDs), which generally emit a broad spectrum of visible, white light and little UV. That may sound like good news for insects, but there is a lot of variability in the proportions of different wavelengths emitted by LEDs, with the 'cool white' types producing lots of short-wavelength, blue light. These types seem to be as, or more, attractive to insects than the old sodium lamps. To complicate matters, different insect groups seem to respond differently too, so that LEDs tend to attract more moths and flies, but fewer beetles, compared to sodium lamps.

There is no doubt that our lights are the cause of both death and confusion in the insect world, but are they really having a significant effect at the population level? As is so often the case, we do not know the answer. It is hard to design satisfactory real-world experiments on the large scale that would be needed with flying insects. Ideally one might have multiple replicate patches of lit and unlit habitat, each carefully monitored over time, but these patches of land would have to be a very long way apart for the light not to spill over into the 'unlit' patches, and skyglow would be a problem unless the experiment could be conducted in a spectacularly remote place such as the middle of the Congo. It would make a fascinating grant application, but I doubt that anyone is likely to fund it anytime soon.

Sticky Cave Glow-Worm

In the caves of New Zealand lives an unusual creature known to scientists as *Arachnocampa luminosa*. It has become quite a tourist attraction, with many thousands of visitors queuing to see the tiny animals, which tend to cling to the cave ceiling, where they glow a gentle bluish-green. In caves where they are abundant they can create the illusion of standing beneath a star-filled sky.

The sticky cave glow-worm is not a worm, and neither is it related to the glow-worms found in Europe, or the fireflies of North America and the tropics, both of which are actually beetles. Instead it belongs to a group of flies known as fungus gnats. Unlike most fungus gnats, which live up to their name by eating fungi, the sticky cave glow-worm is predatory. The larvae – which could also be called maggots, since they are the immature stage of flies – start glowing as soon as they hatch from their egg. They spin themselves a silken nest on the cave ceiling, and then dangle dozens of silk threads down, each adorned with sticky droplets which glisten like a string of pearls in the light coming from the larva. The light attracts small flying insects, which become stuck to the droplets. The glow-worm larva pulls in the thread, eating it as it goes, and then consumes the struggling insect while it is still alive.

13

Invasions

The modern world has seen many organisms redistributed by man from their natural homes to new parts of the globe – sometimes deliberately, but more often by accident. Brown and black rats spread around the world, for example, stowing away in our boats to reach even the most remote oceanic islands. Historically, many deliberate introductions were carried out, sometimes for no good reason. Starlings were introduced to North America in 1890 by the Englishman Eugene Schieffelin, and are now regarded as major pests (while, sadly, their numbers dwindle in the UK). He also unsuccessfully attempted to introduce bullfinches, skylarks, chaffinches and nightingales, apparently because he was a fan of Shakespeare and all these birds are mentioned somewhere in the bard's works. This might seem slightly nuts to us now, but introducing non-native species for fun was quite fashionable in the nineteenth century, with 'acclimatisation societies' in America, Australia and New Zealand actively promoting the practice.

While some species were introduced for whimsical reasons, others were brought in so they could be hunted, for food or entertainment. Rabbits were released in Australia, where they multiplied like, erm, rabbits, with disastrous ecological and economic consequences. Foxes were also introduced so that British expats had an excuse for donning a red hunting jacket and galloping about the landscape with a pack of hounds. The

foxes thrived despite the hounds, helped by the ready availability of rabbits for supper. Had they stuck to eating rabbits
all would be well, but unsurprisingly they quickly developed
a taste for the native wildlife, particularly for ground-dwelling
marsupials such as the greater bilby (now teetering on the edge
of extinction as a result), and the lesser bilby (which went
extinct in the 1950s).

New Zealand lacked anything worthwhile for colonists to
hunt, being free of land-based mammals and the Maori having
already eaten all the giant moa by the time Europeans arrived. To compensate, the European settlers introduced no less
than seven different deer species: red, sika, fallow, white-tailed,
sambar, rusa deer and wapiti, along with chamois from the
Alps and Himalayan tahr (large mountain goats). The red deer
in particular have multiplied to form large herds that cause
considerable environmental damage through overgrazing.

In the twentieth century, many deliberate introductions of
predatory species were carried out with the aim of controlling pests, although often this too did not work out quite as
well as planned. The cane toad is perhaps the most famous
example, introduced to Australia from South America in the
hope that it would consume the insect pests of cultivated sugar
cane. Unfortunately nobody had explained this to the toads,
which decided instead to eat their way through the rest of the
native insect life of Australia, multiplying prodigiously in the
process and hopping all over eastern Australia in a brown tide
of warty skin. There are now reckoned to be more than 200
million in the country.

The rosy wolfsnail provides a less familiar example. It is a
native of southern USA, a fast-moving creature (for a snail) that
chases down and consumes other species of slug and snail. It
was introduced to Hawaii in the 1950s in an attempt to control
giant African land snails, which had themselves been introduced
as food for people (oh, what a tangled web we weave). The
land snails had got out of hand and were consuming crops, but
the rosy wolfsnails did little to control them, instead preferring

to climb up trees in search of the more tasty indigenous tree snails. Within a few years no less than eight native snail species had been driven to extinction.

Although such deliberate introductions of animals are now illegal in many countries, we continue to move bees around the world, as I discuss in chapter 10. We also seem to turn a blind eye to the trade in exotic plants for gardens, which involves large numbers of plants being moved from one country or continent to another. Garden plants can become major invasive weeds: in the UK, *Rhododendron*, Japanese knotweed, giant hogweed and Himalayan balsam are all well-known examples. Their movement also risks the accidental transport of plant diseases or insect pests, such as myrtle rust to Australia. This South American fungal disease attacks and often kills gum trees, bottlebrush trees, tea trees and many other native Australian shrubs, and has been rampaging through natural ecosystems since its arrival in Australia in 2010. Similarly, the box tree moth, an inconspicuous and normally harmless insect from Asia, was accidentally introduced to Europe in 2007 and is now devastating both ornamental box hedging and our rare wild box trees. Once pests or diseases escape from their native land they often have far more devastating effects than in the ecosystem in which they evolved, where their hosts have had millennia to adapt, and there is generally no way of getting rid of them.

Even without the deliberate transport of animals or plants, the scale of global trade in all sorts of goods means that further accidental introductions of non-native species are inevitable. For example, some invasive species have stowed away in shipping containers. The brown marmorated stink bug* is thought to have arrived in the Americas in a shipment of machinery from

---

* The marmorated stink bug belongs to a family of insects – relatives of aphids – known as shield bugs in the UK after their shape, which resembles a heraldic shield, and less endearingly as stink bugs in the Americas after the foul-smelling fluids they exude when disturbed or attacked.

Asia in 1998, and within just fifteen years had somehow made its way throughout the USA. This large, shield-shaped, mottled brown insect sucks the sap of many different crops, particularly apples, apricots and cherries, leaving scars and dimples on the fruits that render them unsaleable. A major pest of fruit trees and some vegetable crops, it now causes some $37 million of crop damage a year. The Asian hornets mentioned earlier are thought to have arrived in France in a consignment of pottery from China, and have since spread across a large portion of Western Europe. Lurid and wildly inaccurate tabloid headlines have suggested that this species poses a threat to humans, which is nonsense, but it is a major predator of large insects. Unfortunately for beekeepers the Asian hornet is particularly partial to snacking on honeybees: if they locate a hive, worker hornets visit over and over again, picking off a bee each time and carrying it back to their nest, so that the honeybee colony slowly dwindles in size.

Other pests have stowed away in consignments of food, aided by the vast global trade in fresh produce. For example, spotted wing drosophila from Japan were first discovered in California in 2008. These tiny flies, just 3 millimetres long, lay their eggs in slightly under-ripe fruit, so that by the time it is ripe it is full of maggots. The species can race through thirteen generations a year, and so has the potential to multiply prodigiously. It prefers cherries, berries, peaches, grapes and other soft-skinned fruits, and in its first year of arrival in the USA it is estimated to have cost the fruit industry an eye-watering $500 million. I have a friend who lives in Davis, California, who used to pick baskets of delicious cherries from the tree in his garden every year. Since the spotted winged drosophilas arrived, however, he has not had a single edible cherry – every single one now contains a mass of wriggling maggots.

You may be wondering what all this has to do with insect declines, since many of these insects are doing rather too well. But of course the changes wrought by invaders can often have major adverse consequences for native wildlife, including insects.

Some invaders, such as Asian hornets and cane toads, are predators of insects. Take the mosquito fish from the Americas, for
instance, which was introduced all over the world to help control
mosquitoes. Mosquito fish consume not just mosquito larvae
but also any other aquatic insect life they can find. Rats and
other introduced rodents in New Zealand are a major threat
to the indigenous giant weta (a huge, slow-moving, flightless
cricket that is now near extinction). Harlequin ladybirds from
Asia have greatly reduced the abundance of native ladybirds in
the UK. Accidentally introduced ant species, such as the fire ant
and Argentine ant, can have profound impacts on native insect
life, particularly on other ants. For example, the Argentine ant,*
which, as you might guess, comes from South America, has invaded southern Europe, USA, Japan, South Africa and Australia,
plus many oceanic islands, even including remote places such
as Easter Island. It more or less entirely exterminates local ant
species as it invades, with knock-on effects on ecosystems such
as reduced seed dispersal (some native ants carry seeds about).
In southern California, a massive drop in native ant numbers is
driving declines in the rare horned lizard, a specialist ant predator
that unfortunately doesn't seem to enjoy eating Argentine ants.†

The spread of alien competitors or predators such as Asian
hornets has a direct impact on other insect life, while the

---

* Argentine ants are unique in forming vast 'mega-colonies' that can extend
for hundreds of kilometres and contain trillions of individuals. This may
be the explanation for their ability to triumph over and exclude other ant
species. In most ants, colonies are highly territorial and fight each other,
often to the death. In a sense, they are literally their own worst enemy. In
contrast, Argentine ants lack genetic variability, and are unable to distinguish
nest-mates from members of neighbouring colonies. As a result, they behave
as if the entire population were a single, more-or-less harmonious unit with
many queens. One mega-colony in Europe stretches 6,000 kilometres along
the Atlantic and Mediterranean coast, from Portugal to Italy.

† You might be wondering why species such as the Argentine ant seem to live
relatively harmoniously with other species in their native range. The answer
is probably partly that they are controlled by predators and pathogens that
they left behind when invading new regions, and partly that, in their native
range, other species have had millennia to adapt to their presence.

spread of plant pests can have indirect impacts on the species associated with the plants under attack. Dutch elm disease is a well-known example. Elms were once among the most common and impressive of trees in the UK, north-western Europe and large parts of North America. The English elm grew to about 45 metres tall, and features in iconic paintings of the English landscape such as Constable's *Salisbury Cathedral from the South-West*. In 1910 signs of elm ill-health began to emerge in continental Europe, with dead branches appearing in the crown of trees. In 1921 a team of Dutch scientists identified the disease as being caused by a fungus originating from Asia, and it is from this that the rather unhelpful name of the disease was coined. The disease was thought to have probably arrived in Europe in shipments of timber, and is spread from tree to tree by various types of native wood-boring beetle. It jumped from Europe to North America in 1928, once again in a shipment of logs. In Europe, the early outbreaks were mild and rarely caused the death of trees, but in 1967 a more virulent strain arrived from Japan via Canadian logs shipped to the UK for boat building. Within a decade, 25 million trees died in the UK alone. I was born in 1965, and I vividly remember the landscape of my childhood as being full of skeletal, dying and dead trees. English elms do survive as hedgerow suckers, for the disease does not attack small plants, but there are very few mature trees left in the UK.

The wholesale loss of major plant species like elms is bound to have had impacts on the associated insect life, for more than a hundred species of insect were associated with elms in the UK. The two best known are the white-letter hairstreak and the large tortoiseshell butterflies, the former now a very scarce species and the latter extinct as a resident breeding species since the 1960s. The plight of insects associated with elms was probably exacerbated by well-intentioned attempts to control the spread of the disease with insecticides aimed at the wood-boring beetles that spread the fungi. In the USA in the 1950s and 1960s DDT and its relative dieldrin were sprayed three times a year onto elms, resulting in the death of large numbers of woodland birds and, although not documented, almost certainly wiping out large numbers of insects.

Now the elm story seems to be repeating itself with ash trees. Ash dieback is also a fungal disease from Asia, where it naturally infects but does little harm to native ash species. The fungus wasn't formally described until 2006, but we think it arrived in Europe in about 1992 when trees with symptoms of dieback were first found in Poland. Even once the disease was well recognised in continental Europe and known to pose a significant threat to ash trees, the UK continued to import ash saplings from Europe until 2012. The first infected trees were recorded that year, close to the sites that had received imported saplings. The government banned further imports and ordered that over 100,000 nursery trees be destroyed, but by that time the fungus was here.

Luckily a small proportion of ash trees, perhaps 5 per cent, seem to be resistant to the disease, so although we may lose most of our ash trees we can at least replant with the offspring of resistant ones. Unfortunately they may then be wiped out by yet another Asian plague which seems to be heading our way, courtesy of a tiny but beautiful metallic green beetle called the emerald ash borer, that since 2002 has devastated ash trees across North America, killing tens of millions of trees. In 2013 the beetle was recorded about 250 kilometres west of Moscow, and as it is spreading at a rate of about 40 kilometres a year there seems every chance it will spread across Europe. I have two lovely mature ash trees in my garden in Sussex. Both now have ash dieback, and over the coming years I will have to watch them slowly die. As with the loss of elms, there will undoubtedly be consequences for wildlife. No fewer than 239 invertebrate species and 548 types of lichens have been found living on ash trees, with twenty-nine of the invertebrates and four of the lichens only occurring on ash, so these organisms will be hit hard. The centre-barred sallow moth, for example, whose caterpillars feed on ash, is already classed as 'vulnerable' in the UK due to an estimated 74 per cent decline in population between 1968 and 2002.

Native insects can also be adversely impacted by the spread of alien invasive plants, which outcompete the host plants on which the insects feed. Over 2.6 million acres of national parks in the USA are badly infested with invasive weeds, particularly grasses such as buffelgrass and reed canarygrass, and also many familiar European wildflowers such as dandelions, knapweed and ox-eye daisy. Kudzu vines from Asia also now smother native forests throughout the southern states of America. Similarly, lantana, a pretty shrub from South America, now occurs in huge stands in the national parks of Eastern Australia. The effects of alien plants tend to be negative, although not always. In a review of studies on the impacts of invasive plants on arthropods (insects, spiders, crustaceans and so on), Andrea Litt of Texas A&M University observed that 48 per cent of studies found fewer arthropod species in areas that were heavily invaded by alien plants, while 17 per cent of studies found increases. Sometimes alien plants and alien pollinators may even form an alliance: lantana and viper's bugloss in Australia are both pollinated mainly by European honeybees, while in Tasmania I have watched Californian tree lupins and European thistles being pollinated by both honeybees and European buff-tailed bumblebees. The alien plants and bees both benefit, but presumably at the expense of whatever native plants and pollinators once lived there.

Through our endless clumsy redistribution of plants, animals and diseases, we therefore risk a massive simplification and homogenisation of the world's fauna and flora. The danger is that we end up with the same tough, robust species everywhere. Native plants become scarce as they are ravaged by foreign pathogens, outcompeted by invasive weeds, or consumed by foreign pests. Native insects decline as their host plants disappear, and as they are depredated by alien predators, infected with foreign diseases or outcompeted by superior competitors. In some parts of the world, such as Hawaii and New Zealand, whole communities of native plants and animals have been swept away and replaced by a mishmash of species from around the world. Further accidental movement of plants and animals is inevitable given the extent of trade around the world, but we could do much more to reduce the risks, following the example of Australia and New Zealand, which both now have rigorous border searches and quarantine regulations.

For me, one of the great joys of travelling is seeing butterflies, birds, bees and flowers I have not seen before; indigenous wildlife that is different wherever you go. This geographic diversity is part of the explanation as to how our planet came to support such a wonderful variety of life. It took many millions of years for evolution to slowly create unique assemblages of plants and animals in each region of our planet, and only a couple of hundred years for us to muddle them up. We cannot undo what we have already done, but with care we could greatly reduce the frequency of future invasions, and so alleviate one of the pressures on our beleaguered biodiversity.

The Elephant Hawk Moth

Caterpillars are tasty morsels, the favoured food for many adult birds attempting to feed their hungry nestlings. As a result, they have evolved a bewildering array of disguises. Many swallowtail caterpillars look remarkably like black and white dollops of bird dropping. Many caterpillars pull off remarkable impressions of sticks, while the caterpillar of the lobster moth resembles an ant when it is small, and a squat spider when it is larger (an odd disguise, since many birds eat spiders). The caterpillars of the emerald moth of Arizona have two different morphs. In the spring they look remarkably like the oak catkins on which they feed, while the second generation, which is found in summer when the catkins have fallen, instead resembles a stick. My favourite is the elephant hawk moth, perhaps because I have fond memories of finding the caterpillars feeding on willowherbs as a child. When fully grown, the large brownish caterpillar looks a little like the trunk of an elephant – not much of a disguise in England, you might think – but when disturbed it puffs up a forward section of its body, expanding a pair of eye spots complete with pupils that give the overall impression of a rearing snake.

14

The Known and Unknown Unknowns

I vividly remember roaring with laughter at Donald Rumsfeld's seemingly incomprehensible blathering in 2002 when questioned about the absence of any evidence linking Saddam Hussein to the supply of weapons of mass destruction: 'We also know there are known unknowns; that is to say, we know there are some things we do not know. But there are also unknown unknowns – the ones we don't know we don't know.' At the time I thought he was a fool, but with hindsight I think he was making a valid and arguably important point. He was making a distinction between things that we don't know are going to happen but could reasonably anticipate, and things that might occur but which we could not have anticipated based on past experience. There are many factors which we know are harming insects (known knowns), and many more that we know about, and might reasonably anticipate to be harming insects, but where good data are lacking (known unknowns). There are also, inevitably, going to be factors we simply haven't thought of, or which are beyond the current state of scientific knowledge, which we may (or may not) one day discover (unknown unknowns).

In case you think that I too am suffering from a case of Rumsfeldian blathering, let me try to give examples. With regard to insect declines, known unknowns might include the effects of new pesticides, or any of the numerous other pollutants produced by humankind, such as heavy metals like mercury released by mining and industrial processes. Each year we manufacture approximately thirty million tons of 144,000 different man-made chemicals for a huge diversity of purposes – pesticides, drugs, flame retardants, plasticisers, anti-fouling paints, preservatives, dyes, and a myriad more – many of which have pervaded the global environment. High levels of polychlorinated biphenyls (PCBs) and polybrominated diphenyl ethers (PBDEs) were recently found in crustaceans (crabs, shrimps and so on) living almost 7 miles down at the bottom of the Marianas Trench (along with piles of plastic bags). For the vast majority of pollutants there have been no studies of their impacts on insects, other wildlife, or for that matter on humans. Those deep-sea crabs may be just fine, despite the PCBs, or they may not; they wouldn't be the most convenient of creatures to study. It seems a pretty good bet that at least some of these chemicals are having impacts on some insects, somewhere. There are simply so many different chemicals that scientists have not yet had time to investigate, and production of new ones far outstrips the capacity of science to keep up. These are chemicals that we know about (at least in part), but we don't yet know whether they are harmful; they are known unknowns.

Another known unknown is the impact of traffic on insect life. It has become common practice to plant wildflowers on road verges and roundabouts, which can look beautiful and potentially provide large areas of flower-rich habitats, linking our towns and cities. On the other hand, there are two obvious downsides to this strategy: first, the insects drawn to these flowers may be struck by passing vehicles, and secondly, they may be harmed by pollutants. Do these flower strips do more harm than good? I once wrote a grant application for funds to study this, but it was turned down (probably correctly, as I'm not sure that the methods I was proposing to use were completely convincing). Thousands of scientific papers on pollinators have been published in recent years, but very few attempt to address this question. We simply don't know for sure what the net effect of such roadside flower strips is, although the answer is likely to be that it varies from road to road. The danger posed by traffic collisions must depend on the speed of the traffic; there is a lovely strip of wild-flowers down the central reservation of Lewes Road going into Brighton, close to the university where I work. The road has a 30 mile-an-hour limit, but usually the traffic is snarled up and virtually stationary, so one imagines that few insects are struck by vehicles. A snail would have a reasonable chance of reaching that flower strip unscathed. On the other hand, there are also lovely patches of flowers adjacent to the nearby A27 dual carriageway, along which a stream of cars roar at 80 miles an hour or more for much of the day. The scarcity of insect 'splats' on windscreens in recent years is often attributed to the lack of insects, but it could also be partly explained by the improved aerodynamics of cars and the increase in traffic volume, so that many fast-moving vehicles are travelling close behind other fast-moving vehicles which will have swept in-sects from their path. The effects of exhaust fumes are also not understood. What harm did leaded petrol do to insects? Lab studies have found that even unleaded petrol fumes impair the ability of bees to learn and remember the scents of rewarding flowers, while diesel fumes degrade the scent of the flowers directly and so make it harder for bees to sniff out the ones they prefer. The bees on the central reservation of the Lewes road may be unlikely to be killed by speeding cars but, bathed in fumes from the near-stationary traffic on both sides, per-haps they struggle to identify the best flowers and gather the nectar and pollen efficiently. Perhaps the food they collect is so polluted that their larvae die; we don't know.

On a related point, the effect of particulate pollution of our atmosphere on insects has not been investigated. Particulate pollution – small particles of dust suspended in the air – comes from vehicle emissions and dust raised from vehicles, from many industrial activities such as power generation, and also from volcanic eruptions and wildfires (whether 'natural' or induced by climate change). Particulates are known to be having profound impacts on human health, causing an estimated 4.2 million premature deaths in 2016 alone. In humans, inhalation of particulates in the air causes strokes, heart disease, obstructive pulmonary disease and cancer, and impairs intelligence, among other things. Insects also have to breathe, but they do it in a different way to us. Instead of having lungs, with air pumped in and out, insects have a series of small holes along their sides, known as spiracles, which connect to a network of branching, ever-smaller, air-filled tubules that snake through the tissues of the insect. They rely largely on oxygen diffusing in, and carbon dioxide diffusing out, although a few larger insects such as hawk moths and bumblebees can pump their body to force air in and out if they have to. It seems intuitively obvious that having particles (of sometimes toxic substances) inside and perhaps blocking these tubules would be harmful, but surprisingly there seems to have been no scientific study of this.

One of the most controversial aspects of particulate pollution of the atmosphere is the deliberate spread of dust via the exhausts of aeroplanes in an attempt to manipulate the weather, either via seeding raindrops or by reflecting sunlight, a technique known as geoengineering. In 2015 I published a small study with Chris Exley of Keele University in which we revealed that UK bumblebees have surprisingly high levels of aluminium in their tissues, enough perhaps to do them harm. In humans, aluminium has been linked to various ailments including Alzheimer's disease. We had no idea where the aluminium in the bees was coming from. Soon after the paper was published I was contacted by several different people who were convinced that this was evidence to support the 'chemtrail' theory, which holds that governments and the aviation industry are complicit in a global conspiracy to manipulate the climate. Their starting point seems to be that contrails, the moisture trails that planes leave behind, dissipate naturally, but that in about 1995 these contrails began to last longer, because they were now full of chemicals introduced by big business or governments to manipulate people or the environment. Exponents of this theory seem to believe that most or all commercial planes are now loaded with tanks of chemicals and fitted with devices to release these chemicals behind the planes as they fly. Aluminium and other chemicals such as sulphuric acid are, they claim, raining down upon the Earth, where they kill insects, trees and perhaps even people.

I reluctantly agreed to meet a chemtrail conspiracy theorist on a visit to Boulder in Colorado. She seemed entirely sane and rational, and very kindly bought me several nice beers while we sat in the garden outside a bar and I studied the photographs she showed me of weird, artificial-looking cloud formations that she said were the result of chemtrails. She pointed to some unhealthy-looking trees with yellowing leaves around the edge of the car park as further proof, but since it was September, when leaves naturally senesce, I didn't find this line of argument enormously convincing. Nevertheless, I was intrigued, and searched the scientific literature. There wasn't much, but I found a paper by two US entomologists, Mark Whiteside and Marvin Herndon, who used circumstantial evidence to argue that coal fly ash (waste produced from coal-fired power stations that contains various toxins including heavy metals) is being used in chemtrails over North America, and that it is a major cause of insect die-offs.

Supporters of the chemtrail theory are generally dismissed as crackpots, and rightly so, because it is absurd to believe that a conspiracy on the scale they describe could possibly be kept quiet. It is not much more plausible than suggesting that the Earth is flat. Nonetheless, geoengineering itself is real. It has been tested on a small scale, and there are plans for bigger experiments. In 2017 Harvard University announced the start of a $20 million study to test small-scale seeding of the upper atmosphere with water, calcium carbonate or aluminium oxide as a means of combating climate change (the experiments were planned for early 2019, just over a year ago at the time of writing, but I cannot yet find any report of what happened).

Some climate scientists are sceptical as to the advisability of this work. As Kevin Trenberth, one of the lead authors for the United Nations Intergovernmental Panel on Climate Change, is reported as saying, 'Geoengineering is not the answer. Cutting incoming solar radiation affects the weather and hydrological cycles. It promotes drought. It destabilises things and could cause wars. The side effects are many, and our models are just not good enough to predict the outcomes.' It seems pretty clear that it is a bad idea but, like many human technologies that could have consequences for all of us (such as the development of artificial intelligence), it is hard to regulate. One small country could, theoretically, alter the climate of the entire world. When the devastating impacts of climate change start to kick in, it is easy to imagine geoengineering being used in a last-ditch, desperate attempt to avoid disaster, but one that might well make things worse rather than better. What impact it might have on insects we can only guess. As Donald would say, this is a known unknown.

There are many other modern technologies that may pose a threat to insects, but where the evidence is either inconclusive or entirely lacking. Electromagnetic fields are created by all electrical circuits, and we know that insects such as ants can detect them. Similarly, the cognitive abilities of honeybees have been found to be impaired by the strong electromagnetic fields created around high-voltage cables. Bees (and perhaps many other insects) use the Earth's magnetic field to help them navigate. In the case of social bees such as bumblebees and honeybees, being able to find their way home accurately is vital, for if they cannot find their nest they will soon die. How might their navigation be affected by the electrical fields created by the high-voltage power lines that bisect the countryside? It seems plausible that such fields might cause considerable disruption of their behaviour, but no one has looked.

While the fixed electromagnetic fields surrounding electrical apparatus are relatively localised, the radiofrequency electromagnetic radiation from telephone masts, wi-fi and mobile phones is more or less everywhere, zapping around and through us all of the time. These waves are low in energy, in contrast to higher frequencies of electromagnetic radiation such as gamma rays and X-rays, which are of course very harmful to living tissue. On the other hand, exposure to radiofrequency radiation is increasing exponentially as technology advances and demands on bandwidth increase. New 5G technology uses electromagnetic radiation of a higher frequency, between 30 and 300 GHz – the same frequencies as produced by your microwave oven, in fact, a bandwidth that has not been used previously for telecommunication. Because waves of this frequency do not travel far, 5G technology will require the installation of hundreds of thousands of small transmitters at regular intervals along our streets, connecting up a projected 100 billion devices globally by 2025. Any part of the body that is very close to a mobile phone is very slightly heated by this radiation. This effect will be stronger close to the new 5G transmitters producing microwaves.

Because I am heavily involved in public engagement over bee declines, giving talks and interviews, I am often contacted by eccentrics with their own theories as to why bees are declining. As with the contrail conspiracy theorists, the beliefs may be supported by little or no convincing evidence, but nonetheless they are passionately held. Some have claimed that diverse ailments of both bees and humans can arise from exposure to the signals produced by mobile phones, including brain tumours and various cancers such as leukaemia. Indeed, I have met people who are convinced that any exposure to electromagnetic radiation such as those emitted by phones or wi-fi makes them feel ill – so called 'electromagnetic hypersensitivity syndrome'. Sad though this is, the scientific evidence to support them does not seem to be strong. Quite a few large-scale epidemiological studies have attempted to assess whether mobile phone use causes cancer, and most have concluded that it probably does not, although some weak associations have been found in some studies. The effect cannot be very strong, or it would have been clearly detected. In 2011 the International Agency for Research on Cancer (IARC), a component of the World Health Organization, commissioned a group of experts in the field to review all available evidence. They concluded that cell phone use is 'possibly carcinogenic to humans', which doesn't really move the debate on.

Much less effort has gone in to understanding possible effects of electromagnetic radiation from telecommunications on insects or other wildlife. One intriguing study found a very strong negative relationship between the strength of electromagnetic radiation and the abundance of house sparrows in different parts of the city of Valladolid in Spain, and the authors suggest that this may be the explanation for the recent profound drop in sparrow numbers in urban areas across Europe. However, this type of correlative study is not conclusive, as there may be other possibly causative factors associated with the strength of electromagnetic radiation; most obviously, places with more electrical gadgets probably have more people, which might scare away the sparrows. Other studies have placed mobile phones in honeybee hives and detected an increase in 'piping' by the worker bees – high-pitched noises usually emitted during disturbance within the hive – but the studies have been criticised because the replication was low and, in any case, having a mobile phone within a beehive is not a particularly realistic scenario.

Of most concern is that there has been so little research into this area. We have rolled out a succession of global telecommunication networks in a huge, unreplicated experiment in which more or less every living being on the planet is being zapped by a rapidly increasing dose of radiofrequency radiation, without us being 100 per cent sure of the consequences. 5G promises to chronically expose every city dweller to a dose of higher-energy microwaves. On social media I have seen video footage purporting to show dozens of dead honeybees on the ground beneath a 5G transmitter, though of course this is not convincing scientific evidence, and it could even have been faked. I have been contacted by people who are convinced that there is no coronavirus pandemic, that COVID-19 is a fiction, and that all its symptoms are due to 5G. Clearly this is nuts, and potentially dangerous if it were to be taken seriously. It confirms a human capacity for delusion that is quite disturbing. Nonetheless, just because some people are crazy does not mean that 5G has no consequences for the health of people or wildlife. It seems to me that it would be sensible to investigate this properly before plunging ahead. 4G technology is pretty amazing already, enabling almost everybody on the planet to download a constant stream of trivia and meaningless prattle at prodigious speed. Is it such a great hardship if it occasionally takes a couple of seconds for your social media feed to refresh? Couldn't we struggle on for a few more years, and do a little more research on the safety, before rolling out 5G?

Before we move on from the murky world of known unknowns, it would be remiss of me not to mention here perhaps the most controversial of them all: that of the risks posed by genetically modified organisms, often abbreviated to GMOs. GMOs are organisms in which we have deliberately altered the DNA, for example by taking a gene from one organism and inserting it into the DNA of another. To some, these are hugely risky 'Frankenstein's monsters', while to others they are a promising technology that we should embrace. Typically, those that campaign against pesticide use are also inclined to be anti-GMOs, regarding them both as dangerous and 'unnatural' technologies. There are certainly genuine concerns, such as the risk that a genetically modified crop might hybridise with a wild relative, allowing the inserted gene to escape into other species. If that gene is, for example, one that confers resistance to a herbicide, then that could make the weed much harder to control. On the other hand, there are potentially great benefits to GM crops: for example, if a crop can be made more tolerant of drought, or more nutritious. If a crop plant could be made more resistant to insect attack, then that might obviate the need for insecticides, with potential benefits for wildlife. Personally,

I do not think we should dismiss GM technology. However, so far most GM crops have been developed by large corporations, with the clear goal of lining their pockets rather than benefiting people or the environment. 'Roundup-ready' crops are a case in point, since they were developed by Monsanto, the company that makes Roundup (a glyphosate-based herbicide). Their introduction has been accompanied by a great increase in Roundup use, which is clearly of detriment to the environment and highly likely to be harming people too (see chapter 8). Additionally, yields from GM crops have generally proved to be little or no better than yields from conventional crops, while the seeds are considerably more expensive for farmers to purchase. In short, I think GM technology might have the potential to be a positive tool if it were in the hands of a benign organisation, but sadly at the moment it is not.

Other genetic technologies of even less certain advisability appear to be on their way. A recent 'horizon scan' of possible future threats to biodiversity flagged up new RNA-based* 'gene-silencing' pesticides. These are sprayed onto crops from where they are intended to be ingested by insect pests, altering their gene expression. Almost any gene in the pest can be effectively 'silenced' in this way, meaning that it is blocked from working. If a gene vital to health or reproduction is blocked, the pest dies or cannot produce offspring. Theoretically, this could enable the production of a vast array of pesticides each targeted at a specific gene in a particular pest species. These would be harmless to other insects so long as they do not have a similar gene. However, since we have only sequenced the genomes of a tiny proportion of all organisms that exist, knowing which ones might have the same gene as the pest is not simple. We would presumably make quite sure that the pesticide would not block expression of our own genes, but could we be certain

---

* RNA stands for ribonucleic acid. RNA is a cousin of DNA (deoxyribonucleic acid), and it can store genetic information in much the same way.

that they would not impact on one or more of the beneficial microbes that live on and in us?

Now, finally, I'd like to tell you about the unknown unknowns, but of course I cannot. If I were to think of one, it would immediately become a known unknown. It seems highly likely that there are many other human activities that impinge upon insect health in ways that we have yet to recognise, for the pace of development and deployment of new technologies far outstrips the ability of scientists to anticipate unintended consequences. There are probably also many natural factors affecting insects that we do not understand; for example, there may be types of disease that we have not discovered. These would be Rumsfeld's 'unknown unknowns', which of course by definition must remain, for the moment, unknown. Still, it seems quite certain that they must exist, lurking ominously in the abyss of our ignorance.

Electrostatic Bumblebees

Bees are the intellectual giants of the insect world, able to navigate long distances, using the Sun and the Earth's electromagnetic field as a compass, memorising positions of landmarks, and learning which flowers provide most rewards and how to efficiently extract them.

In my own research I discovered that bumblebees accidentally leave smelly footprints on flowers, and that bees sniff flowers for the faint whiff of a recent visitor to decide whether it is worth landing or not; flowers that have fresh footprints on them are likely to be empty. Recent research has revealed another bee superpower: they are able to detect and gain useful information from electrostatic charges. As they fly, bees gather a positive charge, just as we do when walking across a carpet. Flowers tend to be negatively charged, so that as a bee approaches, negatively charged pollen tend to jump off the flower and stick to the bee. As the bee lands, its positive charge and the negative one on the flower cancel each other out. This means that recently visited flowers don't just smell of bee footprint, but also have a less negative charge. It turns out that bumblebees can detect this, apparently by the way an electrostatic field makes tiny hairs on their body stand on end. This provides them with an extra clue that the flower is empty, and that it would be a waste of a valuable second or two to land and search for nectar.

Death by a Thousand Cuts

So, which of all of the stressors that insects face in the modern world is the true driver of insect declines? Who dunnit? The answer, of course, is all of them. In the novel *Murder on the Orient Express*, the famous fictional detective Hercule Poirot eventually concludes that the victim was stabbed twelve times by twelve different people; almost the entire cast of the novel are guilty. Insect declines are driven by all of the factors I have described: habitat loss, invasive species, foreign diseases, mixtures of pesticides, climate change, light pollution, and probably other man-made agents we have yet to recognise. There is no single culprit. In Agatha Christie's novel the victim, Ratchett, might possibly have survived being stabbed once or twice, but of course he could not survive being stabbed twelve times. One could argue about which cuts are the deepest, but it wouldn't be a particularly fruitful exercise.

The value of this analogy is limited, for, unlike the murder of one remarkably unpopular train passenger, the demise of insects has taken place over many decades and across the surface of the globe. Different combinations of factors are likely to have affected different insect species at different times and in different places. Importantly, we now know that many of the stressors that harm insects do not act independently of one another. I have already mentioned how certain types of fungicide that are virtually non-toxic to insects on their own can block detoxification mechanisms within insects. If an insect is simultaneously exposed to both the fungicide and an insecticide, the insecticide can be up to 1,000 times more toxic. Similarly, minuscule doses of neonicotinoid insecticides, far too small to directly kill a honeybee, seem to knock out the immune system, so that any viruses present in the bees then rapidly multiply and kill their host. These same insecticides also seem to impair the ability of bees to regulate their own body temperature and that of their colony. Healthy honeybees cool themselves if they are too hot, and warm themselves if too cold, but bees given small doses of insecticide are less able to, so the insecticide might make the bees less able to tolerate heatwaves. Bumblebee nests given low doses of neonicotinoids do a worse job of keeping the temperature of their brood stable. Fungicides and herbicides alter the gut flora of bees, indirectly affecting their health and resistance to disease in complex ways. In all these examples, the effects of the two different stressors together are greater than the sum of the effects of each stressor in isolation.

So far I have mentioned only interactions between pairs of stressors, mainly because this is pretty much all that science can cope with. A good experiment for looking at, say, the impacts of a pesticide on bees, might involve exposing bees or whole colonies to a range of doses. To give a hypothetical example, suppose we want to investigate the effect of new chemical X on bee colonies. We might choose to spike the food of experimental colonies with, say, either 1, 5, 10 or 50 parts per billion of chemical X. We'd need 'replication' – several colonies at each dose – for every colony is slightly different and will respond differently, and statistical analysis of the results is impossible without a minimum of three colonies at each dose. Ideally, to detect small effects, we would have ten colonies at each dose. We'd also need 'control' colonies, given healthy food, free of the chemical. So far we need fifty colonies. If one also wished to look at the effects of a pathogen, and how it interacted with chemical X, that would at a minimum immediately double the size of the experiment – twenty colonies given each pesticide dose, half of them exposed to the pathogen, half disease free. So now we need a hundred colonies. Adding in a third factor would double the size of the experiment again, by which point it would probably become logistically impossible unless the scientist has a vast budget and an army of helpers. Yet, of course, in the real world, insects are bombarded with multiple stressors all the time. Imagine a hoverfly living in farmland; she feeds on the oilseed rape flowers in spring and gets exposed to a mix of pesticides. Once the rape finishes flowering there is nothing much to eat so, hungry and poisoned, she abandons the farm and flies further afield, past the electromagnetic field of a power line to a flowery road verge, dodging traffic and being exposed to diesel fumes as she goes. All the while her immune system may be attempting to fight off a foreign disease she picked up from a contaminated flower, something made harder because her gut flora have been knocked out by the fungicides she has been exposed to. If she manages to lay any eggs they may be depredated by invasive harlequin ladybirds, or killed by a summer heatwave. We have very little idea how all of these stressors might interact, but we should not be surprised if, one way or another, the hoverfly ends up dead or produces few surviving offspring.

The monarch butterfly provides a well-studied example of how multiple, complex and interacting factors can impact on the survival of insects. The precipitous decline of this beautiful and charismatic insect in North America has driven a plethora of scientific studies attempting to pinpoint the cause. A prime candidate is the increased use of the herbicides glyphosate and dicamba, which was enabled by the development of genetically engineered herbicide-resistant maize and soya bean crops. A farmer growing 'normal' crops cannot easily control weeds with herbicides, as he risks also killing his crop. In contrast, a farmer growing GM crops that have been rendered immune to the herbicide can blanket-spray herbicide directly onto the crop, taking out any weeds but leaving the crop unharmed. In this way, it suddenly becomes possible for farmers to grow fields of crops that are almost completely free of weeds. The caterpillars of the butterfly feed exclusively upon milkweeds – formerly common weeds in arable farmland – but now much less common than they were, meaning less food plant for the butterfly.

However, recent research from Stanford University suggests that there is more to it. Milkweed plants defend themselves against herbivores by producing poisonous cardenolides that permeate their leaves and sap. The monarch caterpillars have evolved a tolerance to these chemicals, and store it in their own bodies to make themselves unpalatable to predators. They advertise their foul taste with bright yellow-and-black warning stripes. The cardenolides within the caterpillars serve a second purpose, helping to suppress a single-celled parasite with the unpronounceable name *Ophryocystis elektroscirrha*. Unchecked, the parasite damages the gut of the caterpillar, either killing it or leading to an adult butterfly with deformed wings and little chance of survival. Ecologist Leslie Decker found that milkweeds grown under elevated carbon dioxide levels produced different cardenolides that were less effective against the parasite.

Conservation organisations often recommend homeowners to plant milkweeds in their garden as a means to encourage monarchs. The most commonly cultivated milkweed species is *Asclepias curassavica*, a native of Mexico with higher levels of cardenolides than the North American milkweeds the monarchs normally feed on. These levels are right at the upper limit that the caterpillars can tolerate. When experimentally grown at slightly elevated temperatures, Matt Faldyn of Louisiana State University found that the milkweeds produce even more cardenolides, and become inedible to monarchs. 'There's a Goldilocks zone for these toxins,' said Faldyn in an interview, 'where they're not too toxic but not too weak, either. With climate change, milkweed may pass this tipping point and leave the Goldilocks zone.'

Climate change is also leading the butterflies to spread further north into Canada than they used to, meaning that their autumn flight back to Mexico gets longer every year. Thus climate change is likely to be subtly impacting on monarchs both via effects on food-plant quality and by stretching their annual migration.

Even back at their winter resting grounds, all is not well. In California, twenty of the monarch butterflies' overwintering sites have been damaged by human activity in just the last five years, and others are under threat from housing developments. In the Sierra Madre mountains of Mexico, the overwintering sites face threats from deforestation and mining. One of the great defenders of these sites, the former logger turned conservationist Homero Gómez González, was murdered in mysterious circumstances in January 2020, along with Raúl Hernández Romero, a local butterfly tourist guide.

It is this combination of factors, some obvious, others subtle, that is driving the decline of the monarch. Ecologists studying the butterfly suggest that it may be near a tipping point, where the population falls below a critical threshold and then inevitably collapses to extinction. Like the passenger pigeon, this once common insect might soon be gone for ever.

It has become popular to think of the health of an organism such as a hoverfly, monarch butterfly, honeybee, human – or even a 'superorganism' such as a social insect colony – in terms of their 'resilience', meaning their ability to recover after exposure to a stress or disturbance. All of these entities have mechanisms that attempt to maintain a stable equilibrium, to maintain the status quo. If the human body or a beehive becomes too hot, mechanisms kick in to return it to the optimum: we sweat, and seek shade, while bees fan their nest to suck in cooler air. If our body is short of food, we become hungry and eat more, while a hive with low food stores sends more workers out foraging.

Imagine here that the health of an organism can be visualised as a marble sitting at the bottom of a steep-sided bowl. A stressor pushes the marble from the centre of the bowl, but it quickly returns to the middle. With every stress that is applied, imagine that the bowl becomes shallower, so that the marble is more easily pushed from the middle and is slower to return. Eventually the marble is sitting in a shallow saucer, and even a slight perturbation is enough to send it rolling right off the edge. Every time our body is subjected to a stressor, such as a heatwave, disease, toxin or physical injury, it uses up our energy to recover, leaving us less able to cope with further stresses if they follow soon after (the bowl becomes shallower). If we are poisoned, starving and suffering from infection, a heatwave may be the final straw that finishes us off. The same concept can be applied to whole populations, or even to ecosystems, both of which also tend to show a finite degree of resilience. Harvest some fish from the sea and the population will spring back, but if you take too many the remaining survivors may be so few that the population is not viable and collapses. This will be more likely if the sea is also polluted or vital spawning grounds have been lost. Chop down a few trees from a rainforest and it will recover, with new trees growing to fill the gaps. Chop down all the trees and the soil washes away so that the forest cannot regrow, and instead it is permanently replaced by scrubby, impoverished grassland. Fertiliser pollution of clear, wildlife-rich lakes can cause them to switch more or less permanently to a turbid state with repeated toxic algal blooms and very low biodiversity.

Our planet has coped remarkably well so far with the blizzard of changes we have wrought, but we would be foolish to assume that it will continue to do so. A relatively small proportion of species have actually gone extinct so far, but almost all wild species now exist in numbers that are a fraction of their former abundance, subsisting in degraded and fragmented habitats and subjected to a multitude of ever-changing man-made problems. We do not understand anywhere near enough to be able to predict how much resilience is left in our depleted ecosystems, or how close we are to tipping points beyond which collapse becomes inevitable. In Paul Ehrlich's 'rivets on a plane' analogy, we may be close to the point where the wing falls off.

The Humble Bagworm

Bagworms are a little-known but widespread group of moths, found throughout the world, and named from the caterpillars' habit of making itself a protective case of leaves or twigs, spun together with silk. Keen pond dippers will be familiar with caddisflies, which have the same strategy. Each species of bagworm uses different materials, so many can be identified simply from the case.

Most distinctive of all is the European snailcase bagworm, which constructs a beautiful, helical case from particles of soil and its own excrement, which looks remarkably like a snail shell. Bagworms stick their head out of their case to feed, on leaves or lichens, and when fully grown they pupate within their case. The adult moths do not have working mouth parts, and live just a few days. The males have wings and fly in search of females, while the females are sedentary, either briefly climbing out from their case to mate, or staying inside and allowing the male to stretch his abdomen into the case to mate with her. She lays her eggs inside her case and promptly dies, after what must be considered a very dull life. Since the females never move more than a few centimetres, one might imagine that bagworms would be able to spread only slowly, but for an insect they have in fact two most unusual means of dispersal. Firstly, if a mated female containing eggs is eaten by a bird pecking apart her case, then the tough eggs pass through the bird's digestive system unharmed, just like the seeds of plants such as blackberry, and are deposited in faeces perhaps miles from where the parent was consumed. Secondly, many bagworms take a leaf from the spider's book and 'balloon' in their first larval stage, spinning a silk thread and allowing the wind to carry them aloft.

Part IV

Where Are We Headed?

I have three children, and I am deeply worried that they are set to inherit an impoverished and degraded Earth. Since the Industrial Revolution, parents have been able to look forward, secure in the knowledge that their children would, on average, have a better life than they did. I fear this may no longer be the case. Today, the future is uncertain; there are clear signs that our civilisation is beginning to unravel.

Of course, our twenty-first-century civilisation is profoundly different from those that crumbled before, such as that of the Mesopotamians or Romans. We have technology they could not begin to imagine, from Twitter to nuclear weapons and geoengineering to genetic engineering. Perhaps we can use these technologies to save ourselves, or perhaps they may hasten our downfall. Then again, the Romans also probably thought that they were pretty smart, and probably could not imagine an end to their civilisation, but end it did. I think dark times may be ahead for us too, and that at the heart of this impending cataclysm is the fate of the tiny creatures that live all around us: the insects.

In Part V I will show you my vision of how we might change direction and head towards a better, greener, cleaner world, vibrant with life. But first, indulge me in an exploration of the world our children might inherit if we continue to recklessly exploit our finite Earth ...

16

A View from the Future

I'm tired, struggling to keep my eyes open. It is 3 a.m., but it is still quite warm; one of the muggy, silent nights that have become the norm of late. There are no crickets chirping, or owls to hoot overhead. I sit on an old wooden chair, the rifle across my knees. I could have brought a cushion from inside, but if I was too comfortable I would probably fall asleep.

By the light of the half moon I can make out the patches of vegetables in the raised beds – rows of leeks, parsnips, carrots, beetroot, with Jerusalem artichokes towering more than 2 metres tall, and unruly squash and pumpkin plants sprawling out of the beds, their swollen fruits almost ready to harvest. Beyond them is our small orchard, with apples, pears, peaches and nectarines hanging darkly from the branches. In April and May we worked for weeks, hand-pollinating the flowers, my three grandchildren climbing up like monkeys to pollinate the higher parts of the apple and pear trees, taking care not to crack a branch or break off any flower buds. Unlike some trees, apples only set fruit if the flowers receive pollen from a different apple variety, so we have to carefully collect pollen from the flowers of each tree, brushing it from the anthers into a jam jar, then apply the pollen to the female parts of a tree of a different variety. We use my father's old paint brushes, precious sable-hair brushes that he used to use for his watercolour landscapes back in a different time.

As the year progresses, we do the same for each crop, carefully hand-pollinating the pumpkins, squashes and runner beans. Squashes are easy, as there are only a few female flowers to pollinate, but runner beans are much more fiddly. We leave a few of the root vegetables in the ground each winter to flower the following year – carrots, leeks, parsnips – then hand-pollinate them, and harvest and dry the seeds for sowing the following spring. We have to be organised and efficient, for this is a matter of feast or famine. We focus mainly on growing fruits and vegetables that we can store to eat in winter and early spring, for that is the hungry time. The pumpkins start to go mouldy by the end of January, but we keep eating them until they turn to mush. The apples we store in the loft where they are safe from thieves, and they usually keep until the end of February, though as the winters seem to get warmer every year they keep less well. In any case, we have had few good years for apples of late, for the climate has become too warm for them here in the south of England. The olives, almonds, figs and nectarines do better in our new climate, but we do not have many of them. We should have planned ahead, and planted more thirty years ago in anticipation of the changing climate.

March and April are the hardest months, when last year's crops are finished but most of the spring crops have yet to produce. Purple-sprouting broccoli is great as it flowers at exactly this time, and we supplement it with wild plants, including nettle shoots, dandelion roots, goose grass, chickweed, and any mouldy old veggies left in the store. Young birch and lime tree leaves bulk out the salads. The kids complain, but they are better off than most.

I am getting too old for this night-time vigil. I was born just after the millennium, and I will be eighty next year. My tired bones ache even in warm weather, and much worse in the winter months. I hear a whine as a mosquito passes my ear, and swat at the unseen insect. They are much more numerous than they used to be, one of the few insects that seem to be thriving. There are no bats to eat them – I have not seen one for decades – and the heavy summer rains provide puddles for the mosquitoes to breed in, while the warmth allows them to multiply fast. There have been cases of malaria in the village recently, which spread northwards through Europe and appeared back in England in the 2060s. Every bite could prove fatal, for we do not have access to the drugs that might prevent it.

My father used to remark how ironic it was that most of the useful and beautiful creatures disappeared, while the pests proliferated. House flies form plagues every summer, with no swallows or martins to eat them. Slugs are more common than ever, for the slow worms, hedgehogs and ground beetles that used to keep them in check have gone. Aphids swarm on our vegetables and fruit trees in summer, sometimes wiping out our bean crops entirely, and causing immature fruit to drop from the trees. When I was much younger these pests were eaten by ladybirds, hoverflies, soldier beetles and earwigs. It is always creatures higher up the food chain that disappear first, because they are always more scarce and breed more slowly. Tigers, polar bears and harpy eagles disappeared long before the deer, seals and monkeys they preyed upon. Pests such as aphids, whitefly, slugs, mosquitoes and house flies breed fast, and so can evolve quickly, developing resistance to pesticides, adapting to changing climates. Sadly, the bees and ladybirds could not keep up.

I check my watch once more. Time barely seems to move. My son will take over at 4 a.m., so I do not have too much longer to go.

What changes I have seen. When I was a teenager, I had so much. We all had so much, at least in the western world. I remember supermarkets piled high with food: exotic fruits such as passion fruits, pineapples, mangos and avocados, even kumquats and lychees, flown in from around the world, available on our shelves for twelve months of the year. It seems crazy now. We took it all for granted. Food was so cheap that we bought more than we needed, then threw away much of what we bought when it went mouldy in the fridge. Plastic packaging bulged from wheelie bins, and was taken away and thrown into great holes dug in the ground, along with mountains of dirty nappies and broken plastic toys, where it would all fester for an eternity. I miss pineapples the most – golden-ripe ones from Brazil that dripped sugary juice as you sliced them. And chocolate of course – ah, how I miss chocolate. I have tried to explain how it tasted to my grandchildren, but of course it is impossible. People ate so much rich food there was an epidemic of obesity, and a wave of self-inflicted diabetes around the world. There are few fat people today.

I need to urinate, and so I stiffly push myself up from the chair and hobble over to the compost bin, my knees creaking. I prop the rifle against the side of the bin, a large box made from rough-sawn wood and filled with raked-up leaves, kitchen waste, weeds and any other organic material we can lay our hands on, including the faeces from our composting toilet, and chicken droppings gathered from their pen. There are about a dozen of these heaps dotted around the garden. Urinating on them helps to add nutrients, particularly precious phosphates, and speeds up the composting process. When the petrochemical industry finally collapsed in the '40s, cheap artificial fertilisers became impossible to obtain. We had to switch back to the old ways to fertilise our crops – carefully husbanding all organic matter, breaking it down and returning the nutrients to the soil. For the many farmers who had exhausted their soils and so were entirely dependent on chemical inputs, it became impossible to grow crops, and their fields were abandoned. In October we go out into the nearby wood, or at least what is left of it, to gather sacks of leaves. Our native oak trees, which for millennia following the last ice age had been the commonest tree in Britain, could not cope with the changing and unpredictable climate. Many died in the drought of 2042, and almost all are now dead, their decaying skeletons a prominent feature of the local landscape. Luckily for us, our nearest wood, which starts just a few yards down the lane from our cottage, also contained quite a few sweet chestnuts, which have coped better with the climate. Their nuts are a welcome addition to our diet, and the leaves we gather and compost to supplement our garden soil further, continuing the work that my father began nearly seventy years ago when he bought our cottage and garden. He knew the value of healthy soil, and built up the organic matter content so that the soil in the vegetable patch is deep, rich and black. If he had not, we would not be able to feed the twelve of us who now depend on this 2-acre patch of land.

As I am about to sit down once more, a rustle in the hedge breaks the silence. I hope it is a rabbit. In my father's day the garden was full of rabbits at night, but they are scarce creatures now, prized for their delicious flesh. The squirrels too have mostly been eaten, along with the rats. I raise the rifle, peering down the barrel, but my eyes are not what they once were and I can't see what is making the noise. I cannot afford to miss, as we only have a few dozen cartridges left, and little hope of obtaining more. I bought the .22 rifle when I was in my thirties, as meat became more expensive in the shops. It was for shooting game for the pot, mainly pigeons and rabbits, to supplement our diet. I have looked after it carefully, for it has proved to be one of my most valuable possessions, but once the cartridges are gone it will be useless, except perhaps as a bluff. Most of all, I hope it is not a person crawling through the hedge. The garden is surrounded by dense hedges of hawthorn, and my father supplemented them with sheep-wire fencing topped with barbed wire, but that doesn't stop thieves from sometimes cutting their way through in the dark and stealing our produce.

This was once such a wealthy country, but now people will risk their life for a few potatoes. The signs were there to be seen long before, but things really began to go awry in the '40s. None of us properly understand all of what went wrong. None of us could believe that a global civilisation with such knowledge and technology was collapsing. It should not have come as such a surprise, for civilisations have collapsed before. Indeed, every civilisation ever built has collapsed. For

Romans living at the height of their empire it would have been inconceivable that their vast and efficient civilisation could be over-run by uncouth tribes from the north, and that their mighty cities would become ruins and chaos. History suggests that great civilisations come and go: the Han, Mauryan, Gupta and Mesopotamian empires were all complex, advanced and very sophisticated for their day, yet they all fell apart, and most people aren't aware they ever existed.

Even before I was born, long before, in the 1960s and '70s, scientists began to warn that we were in danger of altering the climate, that we were polluting our soils, rivers and seas, and chopping down the beautiful tropical forests that once teemed with life. In 1992, 1,700 scientists from around the world issued a 'Warning to Humanity'. They explained that humanity was on a collision course for disaster, eroding and degrading vital soils, depleting the ozone layer and polluting the air, felling rainforests, overfishing the seas, creating acid rain, creating polluted 'dead zones' in the oceans, driving species extinct at an unprecedented rate, depleting vital groundwater reserves, and by this time measurably altering the climate. They warned, quite bluntly, that 'A great change in our stewardship of the Earth and the life on it is required, if vast human misery is to be avoided.' They urged that we must cut greenhouse gas emissions and phase out fossil fuels, reduce deforestation, and reverse the trend of collapsing biodiversity.

Governments paid little heed, and neither did the bulk of humanity. Twenty-five years later, in 2017, the scientists repeated their warning, pointing out that humanity had made almost no progress towards reducing the harm the growing human population was inflicting on the planet. This time, the warning was signed by over 20,000 scientists, including my father; we have an old copy of their report somewhere on the bookshelf. While the ozone and acid rain problems had been partially solved, the scientists pointed out that the other problems had all become far worse, and new ones had come to join them. Their new report documented the scale of the worsening crisis in detail. In the twenty-five years since the first report, freshwater resources per person had fallen by 25 per cent; the number of oceanic 'dead zones' had increased by 60 per cent; the number of wild vertebrates had fallen by another 30 per cent; carbon dioxide emissions had risen by about 60 per cent from about 22 gigatons per year to 36 gigatons; the climate had warmed by about half a degree centigrade; the population of methane-emitting ruminant livestock had risen from about 3.2 to 3.9 billion; and the human population had risen from about 5.5 billion to 7.5 billion. They warned that climate change was in danger of becoming a runaway process, and that we had unleashed the sixth mass extinction event on our planet, 65 million years after the fifth such event wiped out the dinosaurs. 'Soon', they wrote, 'it will be too late to shift course away from our failing trajectory, and time is running out. We must recognise, in our day-to-day lives and in our governing institutions, that Earth with all its life is our only home.' No one listened.

In the same year, a group of German entomologists, assisted in small ways by my father, published data showing that insects on German nature reserves had declined in biomass (weight of insects) caught in traps by 76 per cent over a twenty-six-year period to 2016. The authors of this study warned that, if insects were to continue to decline, then ecosystems would begin to unravel, for insects perform a myriad of important roles. The scientists' warning may have gone unheeded, but the media picked up this study and it received global coverage. I was a teenager at the time, and I remember my father giving endless telephone interviews about it for radio and newspaper journalists, patiently explaining why declining insect numbers could be catastrophic for us all. Yet despite all the hot air and ink, neither politicians nor anybody else took any meaningful action.

Why did we fail to act? We humans do not seem to be very good at grasping the big picture. Although we were aware of climate change, extinctions, pollution, soil erosion, deforestation and so on, we were unable to grasp how devastating their combined effects would be. Even the scientists who wrote the two warnings to humanity did not fully understand this. Scientists tend to work in silos, focused on their own discipline. The climate change scientists warned of the impacts of a disrupted climate, biologists talked about the consequences of loss of biodiversity, fisheries scientists warned of depleted fish stocks, ecotoxicologists studied heavy metal poisoning, or plastics pollution, and so on and so on. None of them could fully anticipate that all of these processes were interlinked, with synergies that no-one could predict.

Our failure to avert the crisis could be blamed on a political system that pressured our politicians to focus on the next election, rather than making long-term plans. Many blamed a greed-driven capitalist system that allowed huge multinational companies to garner far more power than politicians or even entire countries, and shape the world to maximise their profits without regard to the human or environmental costs. They were helped by the almost universal faith in endless economic growth, as well as, I think, the assumptions that economic growth and well-being were interlinked, so that life would get better year by year, as it had for so long, so long as the economy continued to grow. I suspect many also believed that technology would solve our problems, that in the future we were destined to jet about the galaxy as in sci-fi movies. If we used up the resources on Earth, we would simply colonise Mars. That we had not even managed to get as far as the Moon since the 1960s ought to have tipped us off that progress on that front was faltering.

Although most scientists warned against it, some tried to fix our ailing climate with geoengineering, by clouding the atmosphere with chemicals intended to reflect sunlight and seed cloud formation. It turned out that the climate was far too complex to control in this way, and their efforts succeeded only in adding to our pollution problems and making the weather even more unpredictable. Machines were developed to take carbon dioxide from the air, but, given the scale of the problem, the amounts they could extract were woefully inadequate. The more obvious low-tech options for carbon capture – large-scale tree planting and looking after the soil – were ignored, while deliberate deforestation was allowed to continue apace, aided by climate change and the increased frequency of forest fires. My father told me that, back in the '20s, scientists even tried to replace our disappearing bees with little flying robots to pollinate crops, but they were expensive and clumsy compared to the real thing, and eventually abandoned. They also tried to genetically engineer honeybees to make them resistant to pesticides, but an unanticipated side-effect was that these 'super bees' proved to be less resistant to disease: they did not last long.

In the early 2020s, the COVID-19 coronoavirus pandemic bankrupted economies and killed far too many people, as a direct result of the trade in wild animals for food and medicine bringing humans and new diseases into intimate contact. Then as the decade wore on, climate change kicked in with a fury that had been predicted but ignored. Hurricanes repeatedly battered the eastern USA and Caribbean, while wildfires ravaged much of Australia, California and the Mediterranean. Even Scandinavian forests began to burn, and peat layers smouldered in the sub-Arctic, belching more greenhouse gases into the atmosphere. Each year millions of people died from the smog produced by these fires, combined with the pollution from factories and vehicles. Climate refugees forced into crowded temporary accommodation provided ideal conditions for further outbreaks of disease.

By the 2030s it was, arguably, too late. The inexorable rise of the oceans began regularly to breach flood defences, aided by heavy rains and storm surges. Devastating floods crippled many of the world's major cities: London, Jakarta, Shanghai, Mumbai, New York, Osaka, Rio de Janeiro and Miami succumbed to the waters, among many others. Weakened by pandemics of disease, economies were unable to cope with the ever-spiralling cost of installing new flood defences, much of them made from concrete, the manufacture of which also released yet more carbon dioxide. Insurance companies were bankrupted by the scale of the disasters, and property insurance became a thing of the past. Entire regions became submerged; large chunks of Bangladesh disappeared under the water, along with the Maldives, most of Florida, and the fens of England.

Climate change became unstoppable, whatever we humans did, because of what are known to scientists as 'positive feedback loops'. Reduced ice cover at the Poles decreased reflection of the sun's energy, leading to more warming, more ice melting, and so on. The thawing of Arctic permafrost released huge quantities of methane that had been trapped in the ground, with methane being a far more potent greenhouse gas than carbon dioxide. Changing weather patterns reduced rainfall on the Amazon, so that the last remaining rainforests of the region withered and died, finally destroying a 55 million-year-old ecosystem, the richest on Earth. As the thin soils the forests had once held together crumbled to dust, they released yet more greenhouse gases.

Most crucially for us, the capacity of the world to feed mankind began to drop. In the '40s a succession of summer droughts in the wheat belt of North America dramatically reduced availability of this staple grain. Meanwhile in Africa the Sahara advanced southwards, driving countless peasant farmers from their land as their crops failed. There were few places for them to go, for temperatures in equatorial Africa were by then so high that human existence had become almost impossible. At the same time, yields of insect-pollinated crops, including everything from almonds, tomatoes and raspberries to coffee and chocolate, began to fall as insect pollinator numbers dwindled all around the world. Crop pest outbreaks became worse, as pests became increasingly resistant to the barrage of pesticides they had been subjected to for decades, and as the warming temperatures allowed them to breed more quickly. The natural enemies of insect pests, predatory insects such as ladybirds, hoverflies, lacewings and carabid beetles, had long since been exterminated in the farmed landscape. In rangelands, animal dung began to accumulate and smother the pastures as dung beetles and dung flies became scarce, unable to cope with the drugs and pesticides pumped into the livestock which ended up in the dung. Without insects to dispose of the dung, the supply of grass fell, and infestations with intestinal worms that spread via eggs in dung became worse.

As if all this weren't enough, many arable farmers found that their soils were increasingly thin and infertile after a hundred years of intensive farming, much of their soil having washed away or oxidised into the atmosphere. The soils that remained were chronically polluted, depleted of earthworms and other small creatures that once helped to keep them healthy. In warmer and dryer regions of the world, such as the Central Valley of California, the wells that had been used for decades for pumping up groundwater to irrigate crops ran dry, while elsewhere major rivers simply stopped flowing in summer due to over-extraction.

In tropical seas, coral reefs turned out to be especially sensitive to temperature rises, which caused them to bleach and die. Before I was born, my parents learned to scuba dive on the Great Barrier Reef off the coast of Australia, and they used to describe the astonishing diversity of colourful creatures they saw. In just one year, 2016, when I was fifteen years old, half of the Great Barrier Reef died. By 2035, almost all the world's coral reefs were dead, removing the major spawning and nursery areas for many fish that were previously harvested for food. In cooler waters, the increasingly desperate search for fish led to industrial trawler fleets flouting government attempts to limit their catch, decimating what remained of the world's great fish stocks. By 2050 there was almost nothing left in the seas but swarms of inedible jellyfish, which proliferated in the absence of the fish.

Perhaps if governments had listened to the evidence and worked together, civilisation may not have been beyond saving even as late as 2035. Sadly, just when mankind really needed to combine resources and expertise to conquer the greatest challenges it had ever faced, it turned its back on reason. Increasing food prices, falling standards of living, rising unemployment and the ever-growing tide of refugees arriving in developed countries led to street riots, protests and the election of extremist politicians. International alliances were abandoned in favour of isolationist, nationalist policies. Countries put their own interests before those of humankind or those of our shared planet. Agreements on fisheries quotas and tackling climate change were torn up, and international aid was withdrawn. Scientists were derided, mistrusted, and their evidence dismissed. Truth was defined by those who shouted loudest, or had the money to buy it. Some said that we had moved into a post-truth world, whatever that meant; the word became so popular that in 2016 the *Oxford English Dictionary* made it 'word of the year'.

The impact of environmental collapse was far more immediate in developing countries. The floods, fires and famines rendered more than one billion people destitute, homeless and desperate. Millions died in famines on a scale that mankind had never imagined possible, while others attempted to flee, creating mass migrations to the cooler north and south. Civil wars and international conflicts broke out, often drawn up along ethnic and religious lines as people searched for someone to blame for their suffering, adopting ever-more extreme and xenophobic doctrines.

In the more crowded countries of the developed world, we had long relied on food imports to feed us. In 2018, the UK grew roughly half the food it needed. By 2040 there were nearly eighty million people living on this crowded island and, with continuing development of farmland for housing, and declining crop yields on the remaining land, we were importing more than 60 per cent of our food. Even after famines began to devastate many developing countries we continued to import their food, for being rich we could afford it while they could not. But as time went on, and food production slumped globally, it became increasingly hard to buy food at any price.

Supermarket shelves began to empty, and families stockpiled what they could. The great refugee camps outside Dover in England, and at almost all of the Mediterranean coastal ports, became a source of bubbling resentment to many. Why should we feed these people, they asked, when we were going hungry ourselves?

The extreme inequality that was a feature of life in the early twenty-first century also fuelled enormous resentment, for while the poor began to go hungry and the homeless became more and more frequent on the streets, the rich were still able to live in great comfort. In the end, though, the source of their wealth was undermined. As the sea rose and the bees disappeared, so share prices fell, hedge funds folded, and banks collapsed. Hyperinflation eventually made money almost worthless, and everyone poor. We had lost sight of the fact that the foundation of our civilisation, and of the economic growth that had so preoccupied our politicians, was a healthy environment. Without bees, soil, dung beetles, worms, clean water and air, one cannot grow food, and without food the economy is nothing.

There was no sudden collapse of our civilisation. It was more a slow unravelling over decades. For a long time we didn't really understand that it was happening, thinking that progress would soon resume. Life expectancy in the UK had risen steadily for 160 years, from just below forty years in 1850 to over eighty years in 2011, but there it stalled. For the first time since records began, life expectancy began to gently fall, starting amongst the more deprived sectors of society, but few paid much heed at the time. Thereafter life expectancy slowly declined over the next few decades as living standards fell and our health service quietly collapsed. Hospitals were crippled by the burden of an ageing population, the epidemic of obesity and related chronic illnesses such as diabetes in the 2020s, and then by repeated outbreaks of antibiotic-resistant bacteria in the 2030s. By the 2040s, schools, hospitals and roads began to fall into disrepair, policemen, nurses and teachers often found that their pay cheques never arrived, and even when they did they could not afford to feed their families. After a thousand years of increasing urbanisation, suddenly people began to abandon the cities, either because of the flooding or because there was just not enough food to support them. Law and order crumbled, and looting began as people scavenged or stole what they could. Starving, the refugees broke out from their camps, adding to the chaos. Eventually the electricity supply began to falter, with power cuts that lasted for hours, then days, until eventually it never came back on. That was a tough year, for we lost our chest freezer full of food.

The water supply lasted longer, but not by much. Without electricity for the pumping stations, I guess it was only a matter of time before the water stopped running. The nearest stream is half a mile away from our house, and in any case it is heavily polluted, so my brothers and I dug a well to try and find cleaner water. We had to go down about 5 metres through the heavy Weald clay until we hit the water-bearing sandstone beneath. It was back-breaking and dangerous work, as we did not know what we were doing, and had no bricks to shore up the walls. Watering our vegetable plants in the long summer droughts is an endless task; I often reminisce to my grandchildren about how we used to have water that sprayed from hoses onto the garden, and sprinklers we could just turn on and leave to magically do the watering all by themselves.

And so I come to be here, peering into the dark, hoping I do not have to threaten to shoot somebody. We have not done so badly. We were lucky, for we live in a fairly quiet area, and we have a little land to grow our food on. An area small enough to defend, just about big enough to feed the three generations of us that now crowd into our cottage. In the last few years things have begun to get easier. In 2050 there were about ten billion people living on Earth, but today in 2080 there must be far fewer, though nobody is counting. Billions died, most of them of starvation, aided by outbreaks of cholera and ty-phoid, the spread of antibiotic-resistant bacteria, malaria and genocidal wars. It is hard now to find out what is happening in the rest of the world, but here we have had fewer intruders in the last few years. The desperate, starving people that once roamed the countryside, scavenging what they could, are now mostly gone, presumably dead.

My old heart skips a beat as I see some movement, but then a surge of relief rushes through me as I realise the shape is too small to be a human. But what is it? Something small and dark lumbers out from the hedge onto the grass. No, surely it couldn't be?

I cannot quite believe it – a hedgehog. I grin stupidly in the darkness. I have not seen a hedgehog since I was a teenager. I thought they had all gone long ago, but here one is, miraculously, bumbling through the undergrowth in search of slugs. Perhaps this is a sign that the world is slowly recovering. I have noticed that the streams seem to be cleaner of late. There are no pesticides and chemical fertilisers any more, no factories belching out fumes. This year I was able to show my grand-daughter her first peacock butterfly – it had been several years since I had seen any butterfly. The tigers, rhinos, pandas, gorillas and elephants are all long gone, now nothing more than mythical beasts, storybook creatures that she will never see, but perhaps, in time, she will live to see the bees return?

The Periodical Cicada

Cicadas are giant, rather ugly relatives of aphids, with bulbous, wide-spaced eyes and large membranous wings, most often to be found perched on tree trunks in warmer climates. Their chunky body, an inch or so long and more than twice that in some tropical species, contains a hollow resonating chamber. This helps the males produce the loudest noise made by any insect, a repetitive zithering or rattling sound up to 110 decibels that can be heard a kilometre or more away, and which is intended to attract a mate. There are many species of cicada, but a few types found in eastern North America, known as periodical cicadas, have adopted a very prolonged lifecycle, with adults only appearing every thirteen or seventeen years, depending on the species. Their offspring live underground, unattractive brown nymphs which suck the sap from the roots of trees and grow extremely slowly. Somehow they keep track of the years while in complete darkness underground, and all emerge in synchrony, within a few days of one another. More than one million individuals can emerge from a single hectare, sometimes in suburban gardens, creating an appalling din that often causes human residents to evacuate the area. The adults live for only a few weeks, so it is soon over for another thirteen or seventeen years. Scientists think that their prolonged and synchronised lifecycle is an unusual way of escaping predation; there is safety in numbers. Insect-eating birds do eat quite a few, but there are so many that most survive. Populations of cicada-eating birds cannot build up, as it will be thirteen or seventeen years before their next feast.

Part V

What Can We Do?

It is not yet too late. Only a small proportion of insects, and more broadly of life on Earth, has gone extinct so far. Undoubtedly more will be lost, for species are going extinct every day, but just as we could halt the juggernaut of climate change within a few decades if we really applied ourselves, so we could halt, and perhaps even begin to reverse, biodiversity loss. Most of the astonishing wildlife with which we share our planet could be saved, for its own sake, and for our descendants to enjoy. Insects in particular can breed relatively quickly, compared to tigers or rhinoceros, and if we only gave them somewhere peaceful to live, and alleviated some of the many pressures we have placed upon them, they could swiftly recover. Since insects are near the base of most food chains, recovery of insects is a prerequisite for recovery of populations of birds, bats, reptiles, amphibians and so on. A vibrant, green, sustainable future, where we live alongside a teeming multitude of other organisms, great and small, is within our grasp.

The first, and perhaps the hardest, step to achieving such a future is to engage with the public – to convince people, one way or another, that insects are important, and that they need our help. People will not take action to help insects unless they care about their fate. Once we have everyone on board, the rest should be easy ...

17

Raising Awareness

Halting and reversing insect declines, or indeed tackling any of the other major environmental threats we face, requires action at many levels, from the general public to farmers, food retailers and other businesses, local authorities and policy makers in government – in other words, action from all of us. It is the combined effects of the harmful actions of all of us that got us into this mess, and it will take a concerted effort from everybody to get us out of it. It is highly challenging to engineer this as, at present, my impression is that a very large proportion of the population is not strongly engaged with environmental issues. Here in Britain, recent elections and the Brexit debate have seen precious little serious discussion of the environment,* despite the compelling evidence that many of the biggest challenges facing humanity in the twenty-first century relate to our unsustainable over-exploitation of our planet's finite resources. Impending water shortages, soil erosion, pollution and the biodiversity crisis ought to be hot topics of conversation around the world, not least because they will have massive impacts on the economy and our health, but they are stubbornly ignored by most. We ostriches are burying our heads in the sand.

* The December 2019 UK election campaign did spark a bidding war between political parties to see which one could promise to plant the most trees, which was welcome but seemed ill-thought-through, tokenistic, and highly unlikely to be properly implemented.

So far are we from fully appreciating the dire plight of the natural world that it is still regarded as a perfectly normal, acceptable hobby to kill animals for fun. Thirty-five million pheasants are reared and released each year in the UK alone, so that a small number of people can enjoy blasting away at these naïve, semi-tame animals.* There are simply too many of us (and soon to be many more) for it to be acceptable to carry on killing animals for amusement. We need to somehow persuade everyone to treat our environment with respect, to teach children growing up that littering, killing, polluting, are just not socially acceptable. How can we do that when the supposedly great and the good slaughter pheasants and grouse just for weekend entertainment?

Of course, I'm not alone in feeling frustrated. In 2017, more than 20,000 concerned scientists from 184 different countries (including myself) signed the 'World Scientists' Warning to Humanity: A Second Notice'. The warning is pretty blunt. 'Especially troubling', it states, among other things, 'is the current trajectory of potentially catastrophic climate change.' 'We have unleashed a mass extinction event,' it goes on: 'the sixth in roughly 540 million years, wherein many current life forms could be annihilated or at least committed to extinction by the end of this century.' Scientists are, for the most part, conservative creatures. That 20,000 of us would put our name to this statement ought to indicate to the world that this is an issue deserving of the full attention of humankind, but most people have not heard this warning, and even fewer have heeded it. On the other hand, the Extinction Rebellion movement is a clear sign that some people, many of them young, are waking up to the fact that their future is being stolen from them. They feel

---

* Rearing birds for shooting is hugely inefficient and harmful to the environment. Around 60 per cent of the pheasants reared, about 21 million birds, are never shot, instead dying of disease, starvation, in road accidents, or being consumed by foxes and other predators, and is likely to be helping to support unnaturally high densities of these predators, with other knock-on effects on our ecosystems.

frustrated and angry because they know that time is running out, and that if we wait until the likes of Greta Thunberg are old enough to reach positions of political influence it will be too late. A new psychological disorder termed 'eco-anxiety' has now been recognised, which afflicts the increasing number of people who worry about the environmental crisis. As Greta Thunberg said recently, 'Adults keep saying, we owe it to the young people to give them hope. But I don't want your hope, I don't want you to be hopeful, I want you to panic.'

In the meantime, though, the large majority of the world's population are paying not the slightest attention, and simply going about their lives as usual. I would guess that in their day-to-day lives more than 90 per cent of the world's population do not think about environmental issues at all. We worry about paying the bills, about our children's education, about how we will look after our ageing parents, or whether our favourite team will be relegated this season. These are all understandable, immediate concerns, much more tangible than the seemingly vague and distant threat posed by cracks in the Antarctic ice sheets, or the possibility that global crop yields might start to fall as nutrient cycles fail, soils erode, the climate changes and pollinator numbers wane. Even those of us who worry deeply about the environment still often drive a car when we might have cycled, or succumb to the temptation to treat the family to a winter break in the sun. Most of us know that we should not fly or drive too much, but the lure of feeling the sun in winter, and the convenience of getting to work or to the shops by car, is just too much to resist. When shopping, we know we ought to buy the organic, free-range chicken portions at £16 a kilo, but the cheap, factory-reared, three-packs-for-a-tenner chicken is so much better value, so long as you focus only on the money and turn a blind eye to the environmental and animal welfare costs. Left to our own devices, we are mostly lazy, self-centred creatures. Many people still casually throw litter from their car window as they drive along, so the margins of our busier roads are strewn with plastics, simply because they can't be bothered to drop it into a bin at the end of their journey. Even these people, who I would cheerfully tar and feather, must have heard about environmental issues, but presumably they just don't care that their children may one day live in a world knee-deep in plastic refuse.

Such people are perhaps beyond hope, along with those few who still actively deny that climate change is man-made, but the large majority are (I hope) decent enough folk who merely don't fully understand the seriousness of the situation. They may be aware enough to recycle, they might be thinking about buying an electric/petrol hybrid next time they trade their car in, but otherwise life goes on as normal and they assume that it always will.

The key challenge is finding ways to engage quickly with this large majority of the population for whom environmental issues currently seem unimportant compared to the more immediate day-to-day issues that life throws at them. I have struggled with how best to do so for many years, and have not arrived at any completely satisfactory conclusions. In my career as a scientist I have written many scientific papers about bees and the causes of their decline, but it long ago became clear to me that this alone was achieving very little, for scientific papers are generally only read by a handful of other academics. So I started writing popular science books about bees, and more broadly about insects, with the aim of trying to reach a broader audience and, ideally, entice in a few of the non-believers. This has been enormously satisfying, but also a little frustrating, since I find that most people who buy them are the ones who already care. Once in a while, someone with no interest in bees might pick up a copy by chance and become interested, but I imagine this is quite a rare event. I give public talks whenever I'm invited, usually about forty a year, to all sorts of groups, including beekeeper associations, wildlife trusts, gardening groups and the University of the Third Age (U3A), and at book and science festivals, but nearly everybody that comes is already interested in insects or the environment. I write articles for magazines, post on social media platforms, give radio and occasional television interviews and so on, but usually I feel I am trapped in a bubble, preaching to the converted, and unable to reach out to those on the outside.

How do we break out of this bubble? I say 'we' because you almost certainly picked up this book because you were already at least a little interested in understanding and combating the looming environmental crisis. Since you've read this far, I'll assume you are now a fellow convert.

Perhaps we should consider our target demographic. Who has the most power to make change? Top of the list must surely be politicians, for a truly green government could, with a little imagination, make profoundly positive changes. Local councillors and governments have considerably less power, but could still do much good if they were so inclined. Sadly, I'm a little cynical about the practicality of influencing our current politicians. Not so long ago I was invited to give a talk in Westminster about the importance of bees. The event was organised by the campaign group 38 Degrees, and I was assured that eighty members of the British parliament (MPs) had promised to turn up. I was excited at what seemed like a great opportunity to influence political movers and shakers. The reality was disappointing: a dozen or so junior staff and one or two MPs actually came and sat through my twenty-minute presentation. The rest came, queued noisily to have their photo taken in front of a large poster of a bee, and then left, paying not the slightest attention to the guy at the front rambling on about insects. How do we persuade such shallow people, who cannot be bothered to spend twenty minutes of their time on learning a little more about it, that the environment should be their top priority? In the UK, the obvious solution is to vote for the Green Party. With our first-past-the-post electoral system this may seem futile, but if enough people voted Green then the bigger parties would notice, and adopt green policies in an attempt to suck up the green vote. Of course, this strategy will only be effective if there are enough green voters, which at present there are not, bringing us back to the central challenge.

Petitions have proved to be a tremendously popular way of attempting to influence policy. In the UK the government has its own petition website, and promises a written response to any petition reaching 10,000 signatures, and a debate in Parliament if it reaches 100,000. Social media are awash with calls to sign petitions, which bounce about in the echo-chamber of Twitter, Instagram and Facebook. I'll happily admit to having promoted many over the years, though of late I have suffered from petition fatigue. I am doubtful whether they have much effect. The written responses from government after 10,000 signatures are usually bland, offering platitudes but no meaningful action. If 100,000 votes are reached the petitioners feel they have achieved a great victory, but the ensuing debate usually involves perhaps a dozen or so poorly informed politicians who gather in a back room of Parliament to blather on for a couple of hours before retiring for a gin and tonic. I once watched one such debate on the environmental risks posed by neonicotinoids (you can watch them on Parliament TV, should you be struggling with insomnia). It was an unedifying experience, since from the outset it was abundantly clear that none of those involved in the debate had more than a rudimentary grasp of the subject – which is complex and technical – and in any case such debates have no power to make any policy decision. I am not suggesting that petitions are a waste of time – they actually take up very little time – but don't expect them to achieve much. There is a danger that people feel that the job is done, just because their favoured petition has reached a certain number of signatures. We will not save the planet simply by signing petitions, no matter how many we sign; they are little more than a displacement activity.

If you are feeling somewhat despondent at this point, let me give you a little hope. There is one inspiring example where a petition has led to real action. In January and February 2019, spurred on by the Krefeld study that had revealed massive declines in German insects, the people of the German state of Bavaria came out en masse to sign a petition. The text of the petition was four pages long, calling for detailed changes to the state's nature protection laws in order to fundamentally change farming in the region, and create a network of insect-friendly habitats. The proposals were radical, calling for a minimum of 30 per cent organic farming, for 13 per cent of the state to be set aside for nature, for 5 metre-wide buffer strips around streams, proper legal protection for all hedgerows and trees, and much more. Unlike the usual UK petitions, which one can sign online from the comfort of one's armchair, the German petition required physical signing, and people queued in the cold winter weather, sometimes for hours, to do so. Quite a few were even dressed as bees, which perhaps helped to keep them warm. Nearly two million people signed, far exceeding the minimum of 10 per cent of the state's registered voters required for the petition to go before the state parliament.

The governing party in Bavaria is the Christian Social Union (CSU), a right-wing, traditional, conservative party with weak credentials on environmental issues. With the support of the farming lobby it attempted to water down the proposals, but the grassroots movement kept up the pressure, and on 3 April the bill was approved. Obviously realising that the best strategy was to embrace the idea, the Christian Social Union's party leader Markus Söder proudly declared it 'the most sweeping nature protection law in the whole of Europe'. He further announced that 100 new government jobs were being created to implement the new laws, along with between €50 and €75 million of funding. Interestingly, Josef Göppel, one of the more environmentally aware politicians within the CSU, said, 'We should rediscover that conserving the diversity of life is what being conservative is all about.' I live in hope that conservative politicians elsewhere might embrace this sentiment.

The developments in Bavaria spurred the national government into action, with the environment minister Svenja Schulze announcing in February 2019 that there would be annual funding of €100 million for insect protection, one-quarter of this to go into research on insect declines. Her plans also include a national ban on glyphosate (the now-notorious herbicide we met in chapter 8, that has been linked to both bee ill-health and elevated risk of cancer in people). Meanwhile, three other German states, Brandenburg, Baden-Württemberg and North Rhine-Westphalia, are planning their own referenda on new measures to support biodiversity. German politicians are now pushing for much of the massive farm subsidies distributed by the Common Agricultural Policy to be directed towards conservation. In Germany at least, there are reasons to be hopeful that their politicians are finally on board.

Back in the UK, it seems we still have a way to go to catch up with Germany. Our petition system does not have the same legal power as in Germany, and in any case the grassroots movement here is not yet strong enough to mobilise signatories in the numbers seen in Bavaria. Likewise in the USA, home of *Silent Spring*, the environmental movement was helpless to prevent the Trump administration from rolling back environmental legislation and slashing the budget for the Environmental Protection Agency.

An alternative way to influence politics we might consider is to become politicians ourselves, but I must confess I have not yet tried this extreme measure. Sadly, few people with training in environmental subjects seem to gravitate into politics. So far as I can ascertain, only twenty-six current MPs in the UK have degrees in science, engineering, technology or medicine (out of 650). Not one has a degree in biology, ecology or environmental science. It is perhaps not surprising that there is little political debate here on environmental issues, and that when it does occur it is rarely well-informed. If you have the stomach for it, and some ecological knowledge, do consider a career in politics.

Getting our politicians on board may be a tough nut to crack, but I would suggest that our next priority demographic, children, are less of a challenge, but only if we get them young.

When I give one of my talks about bees or wildlife gardening in a village hall or local theatre I often find myself looking out over a sea of white hair and bald heads. I would guess that 90 per cent of my typical audience are retirees. I intend no disrespect to the elderly (I'll be one myself soon enough), but we'll all be dead before too long.* If we are really going to change the future it is vital that we engage with children, and somehow encourage them to hang on to the childish enthusiasm that most have for bugs. These are people who have a whole lifetime of decision-making ahead of them, and who might save the world, or at least what is left of it.

Teenage children can be much harder work (and in so many ways). Against my better judgement, on rare occasions I have been persuaded to give talks about bees to groups of secondary school kids, but they are usually an unreceptive audience. Many are looking at their mobile phone, or whispering, or throwing chewed-up balls of paper at one another. My best jokes and most fascinating facts sail past unnoticed like tumbleweed. It doesn't get much better at university level. As part of my duties as an academic, each year I am assigned a group of new undergraduate students, usually eighteen-year-olds fresh out of school. I always take them for a walk on the university campus, which is fairly green and pleasant, with woodland areas, flowery grasslands and several ponds. As we walk I quiz them to get an idea of their background knowledge, waving tree leaves at them, pointing out common birds, and asking which ones they can name. The disturbing outcome is that they usually have little idea what any of these everyday creatures are. Perhaps 50 per cent can identify common British birds such as a robin or a blackbird after some hesitation (but most would also misidentify a jackdaw or starling as a blackbird),

---

* I should acknowledge here the hugely important role that older folk do play in the world of conservation, disproportionately making up the membership of conservation charities, and having time to get involved in volunteering, recording and so on.

while very few can recognise and correctly name a blue tit or wren. Almost none can identify common trees such as sycamore or ash. What is most concerning is that these are all students who have chosen to study ecology at university. I dread to think how sparse the average eighteen-year-old's knowledge of natural history is.

You may wonder why it matters whether people know the names of animals and plants. As Robert Macfarlane argues in *The Lost Words*, a name is more than just a word: it is in a sense a spell that evokes the spirit of the creature it attaches to. If you cannot recognise a brimstone butterfly, you will probably not notice one when it flies by. It does not exist for you because it has no name, and you will not notice or care if it ceases to exist entirely. Controversially, in 2007 and again in 2012 the *Oxford Junior Dictionary* chose to cull many nature words: acorns, ferns, otters, kingfishers, moss, blackberries, bluebells, conkers, magpies, and clover, among others, were all edited from existence, deemed irrelevant to children in the modern world. Even 'cauliflower' was axed, surely still an everyday food, but seemingly judged to be something a child no longer needs to know about. My concern is that a whole generation is growing up for whom nature does not exist, and this is surely something we must resist at all costs.

Primary schools must be the place to start, and these are much more fun to visit than most secondary schools. Young children are, as I have observed before, often naturally drawn to nature, particularly before they hit the teenage years. They are not yet worried about looking cool, and they still have that innate sense of wonder at the natural world that we are all born with but most of us forget as we get older. Take a class of primary school kids out onto the playing field, or better still a wilder area with long grass, bushes and trees, give them some insect nets and pots, and they will scamper about for hours, screaming with squeamish excitement and delight as they attempt to capture slugs, centipedes, earwigs and beetles. It is sad, then, that most children never get this opportunity.

At secondary school, which children in the UK usually start at the age of eleven, ecology and the environment are currently studied a little inside biology classes, but get scant attention and often seem to be unimaginatively taught. 'It seems doubtful that much ecology is taught or learnt in schools,' an academic study of ecology teaching in the UK by P. R. Booth, a member of Her Majesty's Inspectorate of Schools, damningly concluded as long ago as 1979. 'The great majority of sixteen-year-olds have done little or no ecological work, and a large proportion of eighteen-year-olds, even if they have studied A-level biology, have done very little.' Since then the situation has, if anything, deteriorated. The proportion of the biology A-level curriculum devoted to ecology fell just a little from 12 per cent in 1957 to 9.5 per cent in 2017, while the proportion of the curriculum devoted to fieldwork fell from 12 per cent to just 1 per cent over the same period. The little fieldwork students do get often involves exercises that seem to have little point, such as identifying plants in quadrats (rigid wire squares of fixed size) along an ecological gradient. My two elder sons have both studied biology at GCSE and A-level, but to my dismay they were both bored by the ecology part.

What is going wrong? The lack of fieldwork seems to be a fundamental part of the problem. Learning about ecological concepts such as succession, competition, or trophic levels in a classroom is pretty dry, but these things really come to life if they are taught in the field by someone who knows their stuff. Perhaps most significantly, some teachers have little knowledge of ecology themselves, including being unable to identify many common organisms. This leads to a vicious circle of ignorance. Because of their lack of knowledge, teachers may be reluctant to take classes out into nature. In any case, many schools are in cities and do not have ready access to interesting field sites.

A more general problem with ecology teaching is that the subject is complicated and messy. After a lifetime of studying the interactions between insects and plants, it's clear to me that there is much that we don't understand, and often simple experiments do not reach clear conclusions, raising more questions than they answer. This can be seen as part of the fun, but it inevitably makes the subject challenging to teach.

How do we improve the situation, so that children grow up with an appreciation of the beauty, wonder and importance of the natural world, and a basic understanding of the major environmental issues?

In my ideal world, learning about nature would be an integral part of the school curriculum for all children from when they start school at five up to the age of fifteen. From an early age, children would learn about the importance of earthworms, and would do an earthworm survey every year, getting muddy digging a hole and trying to identify the different types of worm they found. They would learn about soil, composting, and nutrient cycles; they would look for tardigrades and rotifers under a microscope; they would pond-dip and catch newts, learn the names of common butterflies and birds, do leaf rubbings and learn about our native trees. Classrooms would have a formicarium (ant house), a wormery, a carnivorous plant or two, and perhaps a fish tank with local pond-life: ramshorn snails, water beetles and dragonfly larvae. Every school would have access to a green, outdoor space where the children could grow plants, perhaps learn how to cultivate some vegetables, and watch bees and butterflies pollinate them. Part of this area should be managed for nature. If such space is not available in the school grounds, local authorities would be charged with locating and designating a suitable patch of land within a short walk of every school. No new school would be built without space for nature.

In this dream world of mine, every school would be paired with a nature-friendly farm, with a small portion of farm subsidies allocated to support the farmers willing to join up to such a scheme and have regular school visits, so that children learn to understand the basics of where the food in their supermarket comes from. They would learn that everything we have, from the oxygen we breathe to the food we eat, depends upon nature, and that we are part of it.

I would advocate that there should be designated classes for nature study in secondary schools, and the option of studying for a formal GCSE qualification in Natural History. The Green Party's only MP, Caroline Lucas, has been calling for the latter for some time. I have already touched upon perhaps the biggest obstacle to having natural history taught formally in schools, which is the lack of sufficient expertise amongst teachers. If we were to go down this route, the government would need to provide support for in-service training for both primary schoolteachers and, at a more specialist level, for secondary schoolteachers. New secondary schoolteachers could be recruited in this area by encouraging graduates in ecological subjects to study for a one-year postgraduate certificate of education. This all requires funding, but surely encouraging future generations to value our planet is worth a little expense?

Aside from politicians and children, who else do we need to bring on board? The answer, of course, is everybody. Gardeners can do a huge amount to help, as could those responsible for management of local authority land, and I will turn to that in the following chapter. Farmers and the food industry I will also return to later. Every human being makes umpteen small decisions every day of their lives that directly or indirectly impact on insects and more generally on our environment, either positively or negatively. We all need to take responsibility for saving our planet. But how can we possibly persuade so many people to care, and so engineer a revolution in behaviour?

This may seem a daunting, and even impossible, goal, but actually I suspect it may be nearer than it might seem. In his bestselling book *The Tipping Point*, Malcolm Gladwell argues that just a few people can change the behaviour of a crowd, that there are tipping points when an idea, belief or behaviour crosses a threshold, tips, and spreads like wildfire. Just as with pyramid-selling, if one person can convince two others, and they each convince two more, and so on, in no time at all huge numbers have converted to the cause, whatever it might be. The emergence of Extinction Rebellion, the rise of veganism,

Jekyll-and-Hyde Locusts

Locusts are simply big grasshoppers. Found in most of the warmer regions of the world, they ordinarily lead innocuous and largely solitary lives. They are camouflaged, usually green or brown, and fairly sedentary, quietly feeding on the leaves of a broad range of plants. They ignore or even avoid others of their kind, except when they are seeking a mate. This all changes if a bout of heavy rains provides favourable conditions for plant growth, which in turn allows the locust population to rise. If young locusts bump into one another regularly, as occurs when populations begin to climb, the tactile stimulation induces a remarkable transformation, both physical and behavioural. The locusts develop bright colours (usually black and yellow), they become more active, and they become gregarious, actively seeking out one another's company, and forming swarms. While the solitary locusts avoid plants containing toxins, these gregarious locusts actively seek out poisonous plants to eat, storing the poisons within them so that they become toxic to predators. They grow fast, and breed fast. Swarms can grow rapidly to contain as many as 200 billion individuals, darkening the skies, with densities reaching 80 million per square kilometre. Under these conditions every scrap of vegetation is consumed within minutes, crops are annihilated, and the swarm moves on. Such plagues have devastated our crops since prehistory: the ancient Egyptians carved images of locusts among their hieroglyphs, and they feature in the Bible and in the Quran. Although locust plagues became less common in the twentieth century, in 2020 huge swarms raged across much of Africa, the Middle East and Asia. With their boom-and-bust strategy, locusts may be one insect that is able to avoid the looming insect 'apocalypse'.

18

Greening our Cities

We often feel helpless in the face of huge, global conservation issues such as climate change, tropical deforestation or the demise of polar bears as the ice melts. Any individual actions we might take seem trivial, too diffuse to make any appreciable difference, and the events are often unfolding in faraway places where we cannot imagine having any direct influence. Thankfully, conserving our insects is something with which everyone can get directly involved, and feel they are making a tangible difference. Unlike polar bears, insects live all around us, in our gardens, city parks, allotments, cemeteries, on road verges, railway cuttings and roundabouts, and these are places that are relatively easy to make more insect-friendly. Gardens alone cover about half a million hectares of the UK, a bigger area than all of our nature reserves, and one that will only expand with proposed house building in the coming years. Our gardens are linked together by these urban green spaces, and our villages, towns and cities are linked by road verges, railway cuttings and embankments. The UK alone has a quarter of a million miles of road verges. There is an opportunity to swiftly turn our cities, towns, villages and gardens into a buzzing network of insect-friendly habitats.*

---

* For much more detail on how we might 'rewild' our gardens and green our urban areas, do read my book *The Garden Jungle*.

The most obvious step that gardeners might take is to plant some pollinator-friendly flowers. It is very easy, and there is an abundance of advice available, albeit not always 100 per cent reliable. Many lists of pollinator-friendly plants have been published. In the UK, for instance, the advice published by the Royal Horticultural Society, available online, is among the most exhaustive and reliable. Garden centres often flag up pollinator-friendly plants with labels, usually adorned with a cartoon bee logo. Broadly, in the temperate northern hemisphere, you won't go far wrong if you fill your garden with old-fashioned cottage-garden plants and herbs, such as lavender, rosemary, marjoram, comfrey, catmint, thyme and hardy geraniums (do not confuse the latter with the related *Pelargonium*, which are useless for our native insects, being adapted for pollination by long-tongued flies found in southern Africa). Squeeze in some native wildflowers too if you have room; in Western Europe, foxgloves, viper's bugloss, or white deadnettle are great choices, but there are many more. Some native plants produce beautiful flowers, and are also the food plant for caterpillars of butterflies: in the UK, bird's foot trefoil and lady's smock, for example, provide food for the caterpillars of common blue and orange tip butterflies, respectively. Avoid annual bedding plants such as busy-lizzies, begonias, petunias and pansies: these plants have been intensively bred to produce large and colourful blooms, but in the process they have often lost their scent or nectar, or the flower shape has been so altered that insects cannot enter, so they tend to be pretty hopeless for insects. Also steer clear of double varieties of flowers such as roses, cherries, hollyhocks and aquilegia, since these are mutants which produce extra petals instead of pollen.

If you only have a tiny garden do not despair, for even a balcony or roof terrace can provide food for pollinators like bees and hoverflies; on the tenth floor of a city-centre building I have seen bumblebees turning up, regular as clockwork, ferrying food back to their nest hidden somewhere in the urban jungle. A few herbs such as marjoram or chives in pots will draw them in, with the added bonus of providing a tasty addition to your cooking.

If you have a lawn, a second, simple step to making your garden into an insect paradise is simply to mow a little less often, saving petrol and your own time. You might be surprised by how many flowers pop up: buttercups, daisies, dandelions, clovers, selfheal and bird's foot trefoil are all commonly found in lawns, but with regular mowing they never get to flower. Ease off for a couple of weeks and flower buds will soon pop up and open, drawing in a crowd of insects.

Of course, to some tidy-minded gardeners these lawn flowers are 'weeds', to be grubbed out manually or sprayed off with herbicide. I have never grasped why some folk are so desperate to have a perfectly uniform, green lawn, unmarred by pretty flowers. The concept of a 'weed' is entirely within our heads; one man's weed is another's beautiful wildflower. If we could somehow engineer a shift in attitude, so that 'weeds' such as daisies or clovers were seen as desirable additions to a lawn, rather than enemies to be battled against, we would save ourselves an awful lot of time, money and stress, while helping nature into the bargain.

Relating to this, you could also make your garden a pesticide-free zone. There is simply no need for pesticides in the garden, and why would you want to bring poisons into the place where your children play? I speak from experience, for I am lucky enough to have a 2-acre garden full of flowers, fruit, vegetables and wildlife, all living in approximate harmony without any artificial chemicals. If you discover a few aphids or whitefly, leave them be, for they are food for lacewings, ladybirds, earwigs, hoverflies and blue tits. Probably they will be eaten soon enough, and if not, it is unlikely they will do too much harm. If you have a plant that is routinely infested with pests then it is a certain sign that the plant is not happy; simply try growing something more suited to the conditions in your garden.

If you would like your garden to be pesticide-free, you should also be wary of the pretty flowers on sale in your local nursery. Sadly, the large majority of garden centre plants, including those labelled as 'bee-friendly', have been treated with insecticides and other pesticides, and the residues are often still present. We discovered this in my lab in 2017 when we screened a selection of UK garden centre plants for pesticides. Ninety-seven per cent of the plants labelled as bee-friendly contained at least one pesticide, with 70 per cent containing neonicotinoid insecticides. The latter are now mostly banned, but I'd be willing to bet that they have been replaced with other insecticides. Far better to buy from an organic nursery (you can find some online), or grow your plants from seed, or plant-swap with friends and neighbours. These options also avoid the environmental cost associated with the peat-based compost that many ornamental plants are grown in, the fertilisers used on them, and the plastic pots they are sold in (most of which are never used again).

While you are at it, you might also write to your local council and ask them to stop spraying pesticides in your local park, and along the road verges and pavements. You might even urge them to phase out pesticides entirely. Thirty years ago, the small town of Hudson in Quebec, Canada (population 5,135), became the first place to ban pesticides. The ban came about thanks to the dedicated efforts of a local doctor named June Irwin, who became convinced that health problems she was seeing in her patients were linked to heavy pesticide use in gardens. She attended every town council meeting for six years to raise this issue, and her dogged determination eventually paid off, with the council introducing a bye-law banning all chemical pesticides within the town limits.

Hudson has since been followed by 170 towns across Canada, including major cities such as Toronto and Vancouver, while eight of the ten Canadian provinces have banned all cosmetic use of pesticides. Thanks to June Irwin, 30 million Canadians now live in pesticide-free areas. Towns elsewhere in the world have followed, from Japan to Belgium and the USA. France has taken the concept to heart, with 900 towns declaring themselves '*villes sans-pesticides*'. That then prompted the government to ban all non-agricultural uses of pesticides nationwide from 2020. Only registered farmers are now allowed to buy them.

The UK has been somewhat slower to pick up on this movement. A number of UK towns and cities have pledged to phase out pesticide use by their local authority, including Brighton, Bristol, Glastonbury and Lewes, and also the London borough of Hammersmith & Fulham, but there seems to be no appetite to restrict domestic use. Surely if the entirety of France can manage without any pesticides in urban areas, we can follow suit? It seems to me that there is little or no downside to doing so, unless you are a manufacturer of pesticides or sell them from your superstore. The upsides, of course, would be more urban biodiversity, and no chance that we or our children would be exposed to these toxins – which include suspected carcinogens – when playing in the garden or in the local park.

There are many other small steps the gardener can take to make their plot a little wilder and a little richer in life. Ponds are wonderful, attracting all sorts of insects such as dragonflies, pond skaters and whirligig beetles, along with amphibians like newts, frogs and toads if you are lucky. Even tiny ponds seethe with life, and provide a place for birds to drink and bathe. Create a compost heap to recycle your own organic waste and you will be giving a home to a myriad tiny creatures, from springtails to tardigrades, millipedes and woodlice, as well as producing your own beautiful rich compost, negating the need to buy compost in plastic sacks from the garden centre. If you have room, sow a wildflower meadow of your own, or plant a flowering tree like an apple, cherry, willow or lime.

Finally, you might try providing homes for some types of insects in your garden. A lot of fairly useless 'bug hotels' are sold in garden centres: for example, butterfly hibernation boxes are commonly sold, but field tests of forty such boxes over two years by scientists at Penn State University in the USA found zero occupancy by butterflies (though quite a few contained spiders). On the other hand, 'bee hotels' that aim to provide homes for solitary bee species often work remarkably well. Bees like red mason bees and leafcutter bees simply need horizontal holes to nest in. You can easily make a hotel for them yourself, by drilling holes of between 6 and 10 millimetres diameter into a block of wood, or by tying bundles of bamboo together. Some fancy commercial designs have windows that allow you to peek inside to see what is happening in the hotels, which is fascinating for adults and a great way of getting children interested. Occupancy of bee hotels is a bit hit-and-miss, depending on whether you have the right species of bee nearby, but it can be high; I often get 100 per cent of the holes occupied. You might also provide a 'hoverfly lagoon', a miniature pond in an old plastic milk bottle or similar: fill it with water and a handful of lawn clippings or leaves, and hopefully it will attract some beautiful hoverflies to lay their eggs.

Of course, in our crowded and urbanised modern world, many people don't have any kind of outdoor space to call their own. If so, you might be feeling frustrated by all this talk of gardening, but you can still get involved. You could follow the example of a group of weekend conservationists based in Stirling, Scotland, who call themselves On the Verge. They spend their weekends digging over any piece of mown, boring grass they can get their hands on, and sowing it with wildflowers – with the permission of the owner, of course. There are now eighty-two patches of wildflowers dotted around Stirling and the neighbouring county of Clackmannanshire, on road verges, roundabouts, parks, school fields, and even one in a prison yard. I had a keen undergraduate student named Lorna Blackmore survey these wildflower plots, comparing them to neighbouring patches of mown grass. Lorna found that the sown patches had twenty-five times more flowers, fifty times more bumblebees, and thirteen times more hoverflies than the mown grass they had replaced. Wouldn't it be amazing if every city and town had a similar scheme, so that all our urban areas were peppered with wildflower patches?

More generally, most local authorities manage an awful lot of land that could be rich in wildlife, but is currently not. If we can persuade them to come on board, there is vast potential. Our parks could have meadow areas, beds planted with pollinator-friendly flowers, wildlife ponds, flowering and fruiting trees, bee hotels and hoverfly lagoons. Every roundabout could become a glorious riot of wildflowers. Cemeteries can be hugely rich in wildlife if they are managed sensitively; some older ones have a diversity of flowers to rival an ancient wildflower meadow, while in contrast others have been strimmed, mown and herbicided into tedious, neat submission.* Local authorities have it within their power to stipulate that all new developments include places for nature, to promote green roofs and tree-planting, and they could protect those brownfield sites that have become rich in wildlife. Our urban areas are often fringed by golf courses, which cover about 2,600 hectares of the UK (Surrey alone has 142). Most golf courses are approximately 50 per cent fairway and green, and 50 per cent rough grassland and woodland, the latter with huge potential for wildlife. Some already support considerable biodiversity, but many are poor, planted with non-native trees and subject to heavy use of pesticides and fertilisers. When granting permission for a new course, local authorities could stipulate that they must be pesticide-free, and that all rough areas must be managed for wildlife, with wildflower meadows and copses of native trees or other native vegetation as befitting the location. Perhaps we could also gently rewild our existing golf courses, taking them back to something more like the ancient golf courses used when golf was first invented in the sixteenth century.

Greening our urban areas would have obvious benefits for insects, for wildflowers, and for the many animals that feed upon insects, but what is perhaps less well understood is how beneficial this might be for us humans. More than a hundred

---

* The UK charity Caring for God's Acre champions looking after wildlife in our churchyards and cemeteries.

years ago Octavia Hill, one of the founders of the National Trust in the UK, said that 'The sight of sky and things growing are fundamental needs, common to all men.' In his book *Biophilia* (1984), the famous American biologist E. O. Wilson argued that humans have an innate instinct to connect emotionally with nature, and that being unable to satisfy that instinct may impact on our well-being. Not long after, a new field of psychological research began to emerge, coined 'ecopsychology' by Californian academic Theodore Roszak, which explores the effect that our ever-diminishing interactions with wild nature have upon our psychological development and well-being. A common contention is that if society is disconnected from nature, then various aspects of an individual's life will be negatively impacted, even to the extent of leading to delusions and insanity. Subsequently, the American author Richard Louv argued in his book *Last Child in the Woods* (2005) that many children growing up in grey, urban environments suffer from 'nature deficit disorder', a suite of behavioural problems brought on by lack of opportunity to play outdoors, in nature. He claims that these problems include attention deficit disorder, anxiety, depression, and a lack of respect for the environment and other life forms. Echoes of these arguments can be found in *Feral* (2013), by the British environmentalist George Monbiot, which promotes the idea that humans have a primal need to experience wild nature.

This all sounds very interesting, and seems like a powerful argument for looking after nature, but where is the evidence? Will we really be anxious, depressed, deluded and insane if we don't get a regular dose of nature, or is this just wishful thinking by environmentalists keen to bolster their cause? Alternatively, is it possible that humans could live perfectly happy, contented lives without ever seeing a blade of grass or hearing the sound of birdsong?

While not every claim made for the benefits of encountering nature stands up to scrutiny, many empirical studies have now been performed, variously by medical researchers, psychologists, social scientists and ecologists, which demonstrate beyond doubt that contact with nature does indeed broadly do us good. Just fifteen minutes spent walking in nature has been found to leave subjects with improved attention and sense of well-being, compared to subjects asked to walk in heavily urbanised areas. Even watching a video of nature gave significant benefits, though not as much as experiencing the real thing. Other studies have found that stress levels of Scottish city dwellers were lower in those living near green space, while in the Netherlands anxiety disorders and depression are less common in those living in urban areas with more parks. In California, people living in areas with more urban tree cover tend to be less fat, and less likely to have diabetes and asthma, having controlled for other factors such as wealth. Expectant mothers living in more green neighbourhoods tend to give birth to heavier infants. Hospital patients with a view of greenery get well more quickly than those with a view of a brick wall. Having a green view from home improves the cognitive functioning of children and the mental well-being of adults. A simulated drive to work through a rural landscape left people better able to cope with subsequent stress in the workplace, compared to those that drove through a virtual urban landscape. Various studies have found that gardeners and allotmenters have a higher level of satisfaction with life, higher self-esteem, better physical and mental health, and less depression and fatigue, compared to non-gardeners. Going on expeditions or camps to wilderness increases multiple measures of mental well-being and connectedness to nature. There are many more, similar studies, but I think I have probably bombarded you with enough examples to make the point: we humans seem to be healthier if we have access to, or sight of, greenery.

As a result of this growing body of evidence, doctors in New Zealand and Australia, and recently in the UK, have started giving 'green prescriptions' to some patients instead of the more traditional drugs. A green prescription usually takes the form of prescribing a regular walk in a park or the countryside, or sometimes taking part in a tree-planting scheme or other outdoor activity. Of course, the exercise itself provides a large portion of the benefit, but the combination with going out into nature appears to be most effective, and more likely to be stuck to by the patient than simply telling them to go to the gym.* In Japan, 'forest bathing' (simply spending time in woodland, with no swimming required!) is commonly recommended by doctors, and appears to have multiple health benefits including boosting immune function.

You might have spotted a flaw in this argument so far. The evidence links human well-being to access to green space, but there is not much known about the quality of that space. Perhaps a boring mown lawn and a *Leylandii* hedge would be enough? Maybe Astroturf and plastic flowers would do the job? Does it soothe our soul and reduce our blood pressure more if there are wildflowers, butterflies or birds? There have been remarkably few studies that have attempted to test whether the quality of green spaces, in terms of their biodiversity, positively affects human health, but those few that have been carried out mostly found that more biodiversity is indeed good for us. Both plant and butterfly diversity of green spaces have been found to have positive effects on measures of human well-being, but bird diversity seems to be the aspect of bio-diversity most strongly linked to human health, especially that of songbirds. Intriguingly, one UK study found that observers got more pleasure from watching the birds in their garden if they could name them, reinforcing the argument that people

---

* Europeans and North Americans have become far less active in the last fifty years, burning on average 500 kcalories fewer per day, a result of more desk-based jobs, driving to work rather than walking or cycling, and using lifts rather than climbing stairs. According to the UK's Policy Studies Institute, in 1971, 80 per cent of seven- and eight-year-olds walked to school, often alone or with their friends, whereas by 1990 fewer than 10 per cent did – and almost all were accompanied by their parents. In the UK, health problems and lost days of work due to inactive lifestyles are estimated to cost the economy about £10 billion per year.

are more likely to care for and empathise with nature if they can recognise it.

One interesting idea is that exposure to a biodiverse environment inoculates us with a more diverse and healthy microbiome, as we call the flora of microbes that live on and in us. Exposure to beneficial microbes during early life has powerful effects on the development of the immune system and reduces the prevalence of chronic inflammatory diseases. Urban dwellers have, on average, a less diverse microbiome, so there does seem to be a plausible link between human health and exposure to microbially diverse environments. There is also some evidence that a high diversity of trees and shrubs provides a more dense canopy, which is better at filtering out air pollution.

Overall, there seem to be immense health benefits associated with providing people with access to green space, and it is likely that these benefits are greater if the area is rich in biodiversity, while knowledge about nature may increase the benefits even further. Inviting nature into our cities and towns would seem to be a simple win-win: good for nature, and good for us. Imagine if every garden was brimming with pollinator-friendly flowers, including native wildflowers, with a mini-meadow, flowering shrubs, a pond, compost heap, bee hotel and a hoverfly lagoon tucked in the corner. This would provide a mosaic of tiny insect nature reserves, which if local councils came on board could be linked by flower-rich road verges and roundabouts, by lines of flowering trees in the streets, by flowering railway embankments, city nature reserves, nature areas in school grounds, city parks and so on, providing a network of interlinked habitats stretching across our crowded country. All new developments would be designed from scratch to maximise biodiversity and public access to green spaces. It seems to me that this is easily within our reach; some of it is happening already, with councils and local authorities banning pesticides and developing plans for helping pollinators, and many a gardener quietly turning their patch into a mini-nature reserve.

Our urban areas could soon become, not just places for people, but places where people and nature live happily and healthily alongside one another, where green leaves and flowers are visible in all directions, where children can grow up surrounded by the familiar buzzing of bumblebees, and where they could learn the names of birds and bees and admire the flashing colours of a butterfly's wings.

Hyperparasitic Wasps

If you know a little about insects you are likely to be familiar with the parasitoids, various wasp and fly species that lay their eggs on or in other insects, and slowly consume them alive, killing them only when they are nearing the end of development.

These parasitoids include the fairy wasps, the smallest of all insects at a minuscule 0.13mm in length, which complete their entire development inside the eggs of other insects. Parasitoids might seem to have it all their own way, but many are attacked by their own parasitoids. For example, the cabbages in my vegetable patch are frequently ravaged by caterpillars of the small white butterfly, which I spend hours picking off by hand. These caterpillars are frequently parasitised by the tiny wasp *Cotesia glomerata*, which injects its eggs into them, and I am always pleased to see the first clusters of yellow wasp cocoons appear next to the cadaver of a caterpillar, showing that help is at hand. However, the *Cotesia* wasps are themselves parasitised by the wasp *Lysibia nana*, which inject their eggs through the butterfly caterpillar and into the wasp larvae within. These *Cotesia* find their host caterpillars using the volatile odours released by the host plant when it is being munched by caterpillars. Remarkably, the odours released by plants being fed on by parasitised caterpillars are subtly different from those released by plants being eaten by healthy caterpillars, enabling *Lysibia* to sniff out caterpillars containing its *Cotesia* hosts.

The Future of Farming

Let us not get too carried away with our grand plans to turn our cities into a giant network of nature reserves. Globally, urban areas cover just 3 per cent of the land, while farmland covers a far larger area at about 40 per cent (with much of the rest comprised of the frozen lands of the polar regions). In the UK, 70 per cent of the country is farmland, and while much of that remains inhospitable to life, the nation's wildlife will always struggle. Most of us seem to have accepted that industrial farming is the only way we can 'feed the world', and we also implicitly seem to accept that wildlife declines are unavoidable collateral damage. In a sense, it is a choice between nature and us; and of course we will always choose us. But is this really the choice? Is it impossible to grow food and support nature at the same time? I would argue that we can do both: that we can have our cake (or carrot) and eat it. I would go further, and argue that if we keep pursuing ever more intensive, industrial farming, with a focus on maximising yields, we will wipe out not just nature but ultimately ourselves, for our very survival depends upon a healthy environment.

At this point, it might be useful to reflect on how we got to where we are, in terms of our modern farming system. A hundred years ago farms tended to be much smaller, comprising many small fields, and with a mix of arable and livestock requiring pasture and hay meadows. Farmers used few or no pesticides and little or no synthetic fertilisers, had much higher farmland biodiversity, but produced much less food. Since 1920 there have been dramatic, sweeping changes. In the UK, for instance, the human population has steadily increased, rising by about 50 per cent from 43 to 66 million, while the number of people working the land has fallen from about 900,000 to fewer than 200,000 today. Eighty per cent of the orchards have gone, as fruit growers found themselves unable to cope with competition from abroad. An estimated half a million kilometres of hedges have also disappeared, as farms have merged and fields have got bigger. Mixtures of synthetic pesticides plus fertilisers are now routinely applied multiple times per year to every field. Livestock numbers have increased, with a doubling in the number of pigs and a quadrupling in the number of poultry, though most are now kept indoors so you may not often see them.

No group of farmers – or politicians, or indeed anyone else – sat down and plotted these changes. All over the world farming has evolved, adapting to market pressures, mechanisation, technological innovation, ever-changing government subsidies and both national and international policies and regulations, the increasing availability of chemical inputs, the emergence of supermarkets with huge buying power, and the public's growing demand for cheap food. Often farmers simply did whatever they thought was necessary to stay afloat. Many small farms have failed, and been swallowed up by larger neighbours. There is nothing to be gained by pointing the finger at farmers; we are all responsible for what has happened to our countryside and where we find ourselves today.

If one looks at the bigger picture, modern farming is part of a staggeringly inefficient, cruel and environmentally damaging food-supply system. Globally, we grow roughly three times as many calories as we need to feed the human population, but about one-third of those calories are wasted, and another third is fed to animals (most of them kept indoors in crowded, inhumane conditions). If we combine the area of pastures used for grazing with that used for growing arable crops that are fed to animals, then three-quarters of all the world's farmland is used to produce meat and dairy products. With the remaining one quarter of farmland we overproduce grains and oils, much of which go to produce unhealthy, carbohydrate- and fat-rich processed foods – pasta, pizza, pastry, cakes, biscuits and so on – while we do not grow enough fruit and veg for everybody in the world to have a healthy diet, even if they could afford it. As a result, we have a global epidemic of obesity and diabetes. If one was designing a system from scratch to feed the world with healthy food in a sustainable, environmentally-friendly way, it would look nothing like our current farming system.

Ideally, what would we want from our food production system? First and foremost, we need to grow enough food so that there is enough for everyone to have a nutritious diet, ensure that it is distributed so that all have access to it, and somehow also make sure that everyone can afford it. Secondly, this system needs to be sustainable indefinitely. It cannot be driving climate change, resulting in deterioration of soils, polluting streams and rivers, or causing declines of pollinators and other wildlife. I have already touched upon the 'sharing-sparing' debate, in which 'sharers' advocate trying to integrate growing food with supporting biodiversity, while 'sparers' argue for farming some areas as intensively as possible to maximise yield so that as much land as possible can be set aside for nature. Our current system is closer to the latter than the former: a high input–high output system that continues to degrade the global environment in a way that is clearly not sustainable. We attempt to conserve nature in isolated pockets of 'spared' land – nature reserves – but nature is still in rapid retreat. The collapse of insect populations on German nature reserves illustrates that this approach is not working, for the spared land is being impacted by the surrounding devastation. Even the most remote spared lands, such as Greenland and Antarctica, are being impacted by climate change.

I also suspect that there is a fundamental flaw in the sparing philosophy. Suppose one were to invent a new wheat variety that gave twice the yield. Would the world's wheat farmers turn half their land over to nature? Of course not. Wheat prices would collapse, and we would find ever-more-wasteful ways of using the surplus, for example by feeding more to animals or using more for biofuels. The farmers would end up farming harder than ever to make ends meet, and nature would not benefit at all.

If instead we consider the sharing route, how might it work? How can we change the direction of our current farming system to make it truly sustainable and supportive of nature, while producing sufficient, healthy food? One option would be to promote and support farmers in adopting a technique known as 'integrated pest management', usually abbreviated to IPM. IPM is really a philosophy rather than a clearly defined approach, with the goal of minimising pesticide use by treating it as the last resort. It emerged as a response to Rachel Carson's *Silent Spring,* developed in the 1970s in the USA via funding from the US Department of Agriculture to various land-grant universities which each developed IPM strategies for different crops. By studying the biology of the pests, encouraging natural enemies, using crop rotations and resistant crop varieties, and various other techniques, the aim is to keep the pest populations low. Only if all this fails and the pests exceed a critical threshold – the point at which they are doing sufficient damage that a pesticide spray becomes cost-effective – does the farmer resort to spraying. A key element of any IPM programme is a practice known as 'scouting', which involves the farmer regularly visiting the crop to count pest numbers. This avoids prophylactic or 'calendar' spraying, ensuring that pesticides are only used when necessary. When I was at university in the 1980s, IPM was considered the gold-standard approach. Use of IPM was made obligatory for all farmers by the EU in 2014 – a bit slow on the uptake, but better late than never – so in theory they should all be using it. Why, then, have we seen a doubling of pesticide applications in the last twenty-five years? The problem is that IPM is poorly defined, so that the EU ruling is impossible to enforce. If challenged, farmers can simply say they are using IPM because they are using one or two elements of IPM, such as rotating their crops. In the meantime, they are bombarded by marketing from the agrochemical companies and their sales reps, encouraging them to use more pesticides. A recent French study of nearly a thousand farms found that most farmers could greatly reduce pesticide use without any loss of yield, and that almost all of them would increase their profit by reducing pesticide use. We are all susceptible to marketing hype, and it appears that farmers have been oversold products they could readily do without, though it may be hard for them to work out which ones those are. It seems to me that a fundamental obstacle to any IPM approach is that minimising pesticide use is the exact opposite of what agrochemical companies desire, and these companies have huge wealth and influence.

Another way we might choose to steer farming is towards sprinkling in a little biodiversity around the edges. For several decades we have explored this approach: in the EU, subsidies are available to farmers to support them in implementing agri-environment schemes such as planting wildflower strips or bird-food strips along field margins, leaving small nesting plots for skylarks in arable fields, and so on (by contrast, in the USA negligible funding is available for such schemes). This approach could complement IPM, since the agri-environment schemes ought to boost crop pollinators and natural enemies of crop pests. In the UK, roughly half a billion pounds is spent each year on such schemes, and there have been some small local successes, but at a national and European scale these measures have not halted the seemingly inexorable decline of our wildlife (though it would presumably have been worse without them). This may be in part because there are simply nowhere near enough of these schemes, but I also suspect that there is a fundamental flaw in the notion of creating areas for nature immediately adjacent to crops that are repeatedly sprayed with pesticides and liberally sprinkled with fertilisers. Sprays drift into the flowers, and pesticides used as seed dressings

contaminate the soil. I would argue that we need more profound change to the way we grow food.

Perhaps a more attractive option would be to encourage
more organic farming, to reduce the pesticide burden on the
environment. Organic comprises a relatively small proportion
of European farming, at 7 per cent of the total farmed area,
with Austria leading the way at 23 per cent and the UK near
the bottom of the list at just 3 per cent. There is clear evidence
that organic farms tend to have healthier soils that store more
carbon, and also that they support more plant, insect, mammal
and bird life than conventional farms, so why not have more
organic farming? A counter argument often used is to point out
that organic farming produces lower yields, so that, the argument
goes, if the world went organic we would need to bring even
more land into production, with negative impacts for wildlife.
The first part of this is undeniably true – organic yields are often
lower – with global estimates suggesting that organic yields are
80–90 per cent of those obtained by conventional farming. On
the other hand, as I have already pointed out, we currently grow
far more food than we need, and we waste approximately a
third of all the food grown in the world, a staggering figure. If
we could significantly reduce food waste, the whole world could
abandon pesticides and we could still easily feed everybody.

Consider, too, that people in developed countries now eat far
more than is good for them. Our excessive food consumption
and poor diet carry enormous hidden costs. At present, 63 per
cent of adults in England are overweight, and 37 per cent are
obese, while nearly a third of children aged two to fifteen are
obese. In the USA the figures are worse; 72 per cent of adults
are overweight, and 40 per cent are obese. UK government
figures estimate that the overall cost of obesity to society (e.g.
via diabetes) is £27 billion per year, and projected to reach
£50 billion by 2050. Similar calculations for the United States
estimate the medical costs of dealing with obesity at $147 bil-
lion, plus another $66 billion in other costs, such as lost work
days and premature deaths.

Not only do we eat too much, but we also eat too much processed food, rich in those cheap grains and oils that we globally overproduce. Many of us also eat far more meat than is healthy for us, or for the environment. Eating grain-fed beef is a spectacularly inefficient way of feeding people, requiring about ten times the land that would be needed if a person ate plants directly, and producing about thirty times the greenhouse gases. Only 3.8 per cent of the protein in plant material eaten by cattle becomes available as animal protein for human consumption. If we could reduce food waste, reduce overconsumption, and switch to eating only small quantities of meat from outdoor-reared animals* (eliminating grain-fed beef entirely), we would need much *less* farmland than at present, while using no pesticides, and we would be much healthier.

This sounds pretty attractive to me, but I think we should go further. Some organic farms look pretty much like conventional farms: they are still trying to grow large monocultures of crops, and are heavily reliant on fossil fuels to power their large machinery. Large-scale monocultures are breeding grounds for pests: even on an organic farm, a large field of wheat does not have high biodiversity, so there are few natural enemies to control outbreaks of pests and diseases. I think there are better ways to grow food, and I would argue that farming could learn something from allotments.† Allotments typically have lots of different crops grown in small patches, and often look quite messy. You might think them an unpromising model

---

* Although some advocate a vegetarian or vegan diet, one can make a pretty good argument for including small amounts of meat in an omnivorous diet, both on health grounds and because small numbers of outdoor livestock can be valuable in sustainable, low-input farming systems. The dung they produce is an important source of nutrients for organic farmers, and their grazing action can be an important management tool for boosting biodiversity.

† For those unfamiliar with the term, allotments are small patches of land, usually available for a low annual rent, and intended to provide space for vegetable and fruit growing for those who do not have access to a sizeable garden of their own. They are popular in many European countries. In North America, similar schemes are usually known as 'community gardens'.

for what food production could look like, but let me tell you a little more about them.

First, a recent study from Bristol University, based on data collected from around the UK, found that allotments had the highest insect diversity of any urban habitat – higher than gardens, cemeteries or city parks, higher even than city nature reserves. Allotments teem with life, probably a result of their higgledy-piggledy nature, with a huge diversity of crops and flowers being grown, some fallow and weedy plots, old decaying sheds, fruit trees and currant bushes, compost heaps, occasional ponds and more. This is also helped by the generally low use of pesticides: Beth Nicholls, who was previously involved in the studies of pesticides in bee food described in Chapter 7, has more recently been surveying pesticide use on allotments near Brighton, and found that most use few or no pesticides. When growing food for our personal consumption or to feed our kids, most of us are much more sparing in the use of pesticides than any conventional farm.

Secondly, Beth has also been working with allotmenters to collect information on the productivity of allotments, and the results are astonishing. Many produce the equivalent of 20 tons of food per hectare (a typical allotment is one-fortieth of a hectare), with a few producing the equivalent of 35 tons or more. This compares very favourably to the main arable farm-land crops in the UK, wheat and oilseed rape, which produce about 8 and 3.5 tons per hectare respectively (much of this going to animal feed or to produce the heavily processed foods that help make us fat). Bear in mind also that the allotment produce is negligible food miles, zero packaging, healthy fruit and veg, often produced with minimal chemical inputs.

Thirdly, research has found that allotment soils tend to be healthier than farmland soils, with more worms and higher organic carbon content, helping to tackle climate change.

Fourthly, a study in the Netherlands found that allotmenters tend to be healthier than neighbours without allotments, particularly in old age. The researchers could not discern whether this was due to the consumption of fresh fruit and veg, the exercise gained from allotmenting, or perhaps the social benefits of having an allotment. Given the abundant evidence that being active, outdoors and in green space is good for us both in terms of mental and physical health, this result is not surprising.

To summarise, allotments seem able to produce lots of food, support high biodiversity, have healthy soil, and make people healthier. This seems like a win-win-win-win situation. Clearly there does not have to be a trade-off between food production and looking after nature.

It is sad, then, that an estimated 90,000 people are on waiting lists for an allotment in the UK. Given the benefits, wouldn't government be wise to free up more land to accommodate these people? Perhaps a tiny fraction of the £3.5 billion currently given out in farm subsidies in the UK could be diverted to purchase land for allotments? Some might also be spent on encouraging even more people to try to grow their own food (either in allotments or in their own garden), perhaps via a programme of public education as to the benefits, and provision of training, support, and free vegetable seeds? Some politicians are advocating a move to a four-day week in the near future. With the extra leisure time, perhaps more people might be amenable to growing their own fruit and veg.

In the UK we currently consume about 6.9 million tons of fruit and veg per year, of which 77 per cent is imported at a cost of £9.2 billion. These are shocking statistics when one considers that our climate and soils are well suited to growing many of these crops. Why do we import two-thirds of the apples we eat, for instance, when we live in a land that is near-perfect for growing apples?* Why do I see leeks from Chile – some 7,500 miles away – on sale in my local supermarket

---

* You may be thinking that it is impossible to have British-grown apples in, say, April, but with the right variety and use of modern storage techniques it is perfectly possible to have crunchy home-grown apples for twelve months of the year.

in March, a time of year when home-grown leeks are readily available? Very crudely, under allotment-style management, all of the fruit and veg we currently consume could be grown in the UK on just 200,000 ha of land (the equivalent of 40 per cent of the current area of gardens, or just 2 per cent of the current area of farmland).

Of course, we could not grow avocados, bananas, or much of the other exotic produce now routinely on sale in our supermarkets year round, and there are also limits to the seasonal availability of home-grown produce, but we could be far nearer to self-sufficiency than we are at present. We could get closer still if we could all learn to value local, fresh, seasonal produce more, changing what we eat throughout the year to reflect the natural cycle of availability as we once did. We will always have to import some produce, but so long as this is largely of products that keep long enough to be transported over land or sea, the associated carbon costs are relatively small.* We could offset imports against export of the crops that grow well in our climate, such as strawberries, potatoes, cherries and peas, so that overall we would not be net importers of fruit and veg.

It is worth examining what it is about allotments and small-scale veggie patches in gardens that enables them to produce abundant food while supporting a healthy, biodiverse environment. There are a number of factors. Small patches of crops, or intermingled rows of different crops, are much less susceptible to pests, which find it much harder to locate their preferred food amidst all the other foliage. Different crops are harvested at different times, so the allotment is never stripped bare – as happens when an arable crop is harvested – so the soil is not left exposed to erosion, and organic matter can build up over time. The latter is helped by the addition of home-made

---

* While shipping food over land or sea adds relatively little to the carbon costs associated with the produce, transporting food by aeroplane (for example, grapes from South Africa) is far worse for the environment and is something we should be striving to reduce or eliminate.

compost, for almost every allotment has a compost heap. The roots of perennial crops such as fruit bushes, rhubarb and trees also help to hold the soil together. Natural enemies of crop pests, like ladybirds, ground beetles and hoverflies, tend to be much more abundant, and the diversity of vegetation provides plenty of places for them to hide, so even if pests do find a crop they tend not to flourish for long. As a result, it is easier to grow plentiful fruit and veg without using pesticides. Pollinator populations are high, also benefiting from the diversity of habitats, so crop yields are not limited by a shortage of pollinators. By growing dozens of different crops in close proximity, the allotmenter or veggie gardener gets multiple harvests per year, rather than just one, which helps to increase the overall annual yield. Different crops can be grown next to one another, making maximum use of the space in a way that is much more akin to a natural plant community than is a large monoculture.

Even in my wildest dreams I do not imagine that everyone is going to wake up tomorrow and want to take up allotmenting or dig over half their garden for veg growing, so there will always be a need for commercial fruit and veg production. As with the main arable crops, such as wheat and oilseed rape, most commercial fruit and veg in the UK are normally grown in large-scale monocultures, but this does not need to be the case. There are commercial farming systems which resemble scaled-up allotmenting: permaculture, agroforestry, and biodynamic farming are all variants on this theme. They are sometimes regarded as alternative, left-field, 'hippy' approaches to food production, but their basic tenets are ecologically sound. All three place an emphasis on regenerating the soil, and building up organic matter and populations of soil creatures such as worms. All involve using a diversity of crops, including perennials and annuals, so that there are no large monocultures.

Agroforestry is simply the practice of growing trees or other woody, perennial plants in close proximity to annual crops, and has been practised in various forms for thousands of years. At its simplest, it might involve growing lines of productive fruit trees on pasture which is used for grazing animals or for free-range hens to roam. Depending on what trees are planted, they might provide multiple benefits, including an edible harvest such as fruits or nuts, shade for livestock or for shade-loving crops, fire-wood or building materials, browse for the animals, mulch for other crops, improved drainage, reducing flooding and holding the soil together. Some tree species might be included because they fix nitrogen from the atmosphere, boosting soil fertility. In the tropics, coffee is commonly grown as a monoculture, with the attendant problems of soil erosion and high pest pressure necessitating high levels of pesticide use. Coffee is naturally a shrub that thrives in shade, and it can be grown much more sustainably using taller rainforest trees to provide that shade. This approach vastly increases the numbers of species of wild birds, mammals and insects living in the plantations, reduces the pest pressure, suppresses weeds, and provides more reliable pollination for the crop. This 'shade-grown coffee' also sells at a premium because of the significant environmental benefits it provides.

Permaculture is a little harder to explain, and is to my mind a little woollier: it seems to me it is more of a philosophy than a science, focused on working with nature rather than against it – something I would of course wholeheartedly endorse. It was invented in the 1970s by Bill Mollison, a scientist at University of Tasmania, along with his PhD student David Holmgren. Mollison was inspired by watching marsupials grazing in the lush temperate Tasmanian rainforests, and his notion was to build environments in which humans could live as part of functionally complex and interconnected, sustainable, living systems. He advocated protracted and thoughtful observation of the interactions and functions of organisms in any particular area before attempting to 'consciously design landscapes which mimic the patterns and relationships found in nature, while yielding an abundance of food, fibre and energy for provision of local needs'. In practice, the schemes Mollison came up with involved growing multiple useful plants together, from trees and shrubs to herbs and fungi, along with encouraging both wild and domestic animals. I can't make up my mind if Mollison was a visionary genius or a mad hippy, or both, but clearly his heart was in the right place.

Biodynamic farming is a concept developed by Rudolf Steiner, an Austrian social reformer, back in the 1920s, and essentially a negative reaction to the very early days of chemical agriculture. Steiner was concerned at the apparent deterioration in the health of crops and livestock, which he attributed to the growing use of artificial fertilisers. Biodynamic farming has much in common with organic farming: prohibiting the use of pesticides and promoting many very sensible practices including crop rotations, setting aside 10 per cent of the farm for nature, and generally looking after the land and producing healthy food.

However, there are aspects to biodynamic farming that are beyond the boundaries of conventional science. Biodynamic farmers create what they call 'preparations' from, for example, crushed quartz rock stuffed into a cow's horn and buried in the ground, or by stuffing a deer's bladder with camomile flowers. These preparations are either added to compost or sprayed onto the land in essentially homeopathic quantities. Eccentric as this may all sound, I was recently lucky enough to be given a tour of Plaw Hatch biodynamic farm in West Sussex, a farm run by a like-minded community of folk, many of whom live on site. Over a communal lunch which started with a blessing of the food, I challenged the staff about the scientific basis for their practices. I was a bit nervous that I might upset them, and one or two of them were a little defensive and quick to express their belief that their preparations were effective. What was more interesting was that several others said they weren't sure whether the preparations worked, but that as far as they were concerned it didn't matter. The preparations are made communally as a social exercise once or twice a year, bringing the staff on the farm together to pick the flowers in a kind of bonding or team-building exercise. There's no scientific evidence that the preparations work, but equally I can find no evidence that they don't. I'd be curious to do some proper experiments to test them, but I imagine that most funding bodies would not take a request for funds seriously. On the other hand, even if these preparations do nothing to improve crop growth, and their only role is to promote social cohesion among the group, that is surely more than enough. After all, conventional companies often spend a fortune on sending their staff on team-building exercises.

Plaw Hatch is a truly mixed farm spread over about 80 hectares, with outdoor chickens, sheep, pigs and cattle, some grain crops, pasture, and a huge diversity of vegetable, fruits and flowers grown for cutting. It has its own dairy, making a range of cheeses and yoghurts, and a farm shop where the community sell all their produce direct to the public. Almost all their food is consumed by people that live nearby, and by the farm staff themselves. The one hectare of kitchen garden produces a little over 20 tons of fruit and veg per year, despite quite a bit of this area being taken up with the flowers grown for cutting.

As I wandered round with Tali, my guide, we saw an abundance of butterflies and bees buzzing about, particularly among the fruits and vegetable crops. Tali was keen to find out how they could encourage even more insects, and I was happy to make some suggestions, but it seemed to me that they were doing pretty well already.

One might argue that the downside of these types of 'alternative' farming systems is that they tend to be much more labour-intensive. Plaw Hatch employs about twenty-five people, whereas the national average for a farm of this size would be about 1.7 people. Industrial farming is heavily mechanised, and so requires very few people (which of course is a major driver of the demise of rural communities). To scale up the extent of biodynamic or permaculture farming to provide a significant proportion of our food supply would require getting many more people back onto the land. Would that, however, be such a bad thing? It is predicted that many traditional occupations will disappear in the next few years as advances in technology and artificial intelligence make us humans redundant. Perhaps one way we could find gainful employment is by expanding small-scale agriculture?

It is obviously not possible to redesign our broken global food system without buy-in from every country, but we could make a start locally. Imagine if our cities were scattered through and ringed by allotments, and by small, productive, labour-intensive market garden/permaculture/biodynamic farms, so that most of the fruit, vegetables, eggs and chicken eaten by city dwellers was grown or reared within a few miles of where they live. In the UK, in rural areas with fertile soils suitable for arable crops – East Anglia and parts of the Midlands, for example – cereals and oilseeds would be grown using organic methods or properly implemented IPM, with greatly reduced pesticide inputs overall, and longer crop rotations incorporating fallow years (when the field is left uncropped) or planted with nitrogen-fixing clover to restore soil health. Large fields would be divided by lines of native trees, capturing carbon and protecting the soil. Thriving farmer's markets and veggie box schemes would bring produce from local farms into the cities. In this imagined world, people would have reconnected with nature and with the benefits of eating healthy, high quality, fresh, seasonal, local produce.

But how might we bring about such changes? One obstacle in the European Union has long been the Common Agricultural Policy (CAP) legislation that was first introduced in 1962, with the goal of ensuring high food production and a thriving farming industry across the six member states at the time (France, Germany, Italy and the Low Countries). As the European Union expanded to comprise twenty-eight states by 2019, so the Common Agricultural Policy has effectively driven the intensification of farming across Europe over the last fifty years, via a subsidy system that ensures the most money goes to the biggest farmers, driving small farmers into bankruptcy. It has focused on maximising yields regardless of environmental costs, and it has at times resulted in massive overproduction of food. The Common Agricultural Policy has also impacted on farmers in developing countries, who have to compete to sell their produce against competition from the cheap, subsidised produce offloaded from Europe.

With much upheaval and controversy, the UK recently left the EU. Whatever one's view on Brexit, it has freed the UK from the Common Agricultural Policy, and provides a golden opportunity to turn farming on its head, to make the radical changes that are urgently needed before most of our wildlife and our soils have gone. The £3.5 billion a year* in farm subsidies currently takes taxpayers' money and uses it to support an industrial farming system that produces copious greenhouse gases, damages the soil, overgrazes the uplands, employs few people, pollutes rivers with fertilisers and pesticides, drives wildlife declines, and overproduces unhealthy foodstuffs while under-producing food that is good for us. Why exactly should we pay our hard-earned taxes to subsidise all of this? However, the existence of this subsidy system does mean there is a mechanism already in place that could be used to steer farming in a different direction. Imagine if, instead, this money was given to small-scale, truly sustainable farming systems such as organic or biodynamic farms, aimed at producing food for local consumption, so that such small farms became more financially viable and proliferated. This could be easily done by giving a premium to farmers for not using pesticides, and by giving payments on a sliding scale so that small farmers received disproportionately more subsidy per hectare of land, with a cap on the maximum allowed. At present the average subsidy per farm is about £28,000 per year, but some of the larger farms receive in excess of £300,000 per year.

---

* This subsidy of about £3.5 billion per year to UK farming remained relatively stable for the last few years prior to Brexit, but it is likely to change in the future as the UK develops its own farming policy.

Of course, it would be better still if such changes were made across Europe, rather than just in the UK. If we had remained in the EU we could have pushed for such changes, though getting twenty-seven member states to agree would have been a huge challenge, unless there was a major swing in public opinion.

An alternative, or complementary, approach to rejigging the subsidy system that should be considered by governments everywhere would be to introduce pesticide and fertiliser taxes. Both pesticides and fertilisers pollute and damage the environment, and it seems reasonable that the farmer using them should pay for this. For example, removing metaldehyde (the chemical in slug pellets) from our drinking water supplies costs our water companies many millions of pounds per year, which is currently paid for by the public in their water bills. Norway and Denmark have introduced pesticide taxes, levied at the point of sale of pesticides to farmers, and these have successfully reduced pesticide use. The Danish system is based on applying higher taxes to the more toxic and persistent chemicals, which seems eminently reasonable.

If taxes on agrochemicals were levied, these could be used to support research and development into sustainable farming systems. The current levels of crop yield attained by intensive farming are the result of many decades of heavy investment on research into new crop varieties, growing techniques, the development of new pesticides and technology to apply them. In contrast, there has been minimal investment in research into organic or other alternative farming systems. The UK once had many government-funded experimental farms which researched how best to grow food, and in 1946, in the immediate aftermath of the war, the government set up the Agricultural Development Advisory Service (ADAS) to provide advice and support to farmers. Since then, almost all the experimental farms have been sold off, and ADAS was run down and eventually privatised in 1997. Now, the main investors in farming research and development are the big pesticide companies and other agro-industries. The main advice available to farmers now is from agronomists who mostly work for pesticide companies (some are independent, but nonetheless their main source of information is the marketing and promotion of products by the agrochemical companies). Given that food production is essential to our survival, and that the way we do it has profound impacts on our environment, surely it is worth some investment of public funds in getting it right? We should reinstate government-funded experimental farms, doing research into how to optimise truly sustainable agriculture, and on how to reduce pesticide use in conventional farming. If a gardener or allotmenter can get 35 tons of food from a hectare of land without any training or research and development to back them up, imagine what might be possible if we took a scientific approach to properly evaluating the best practices. Researchers could investigate which combinations of crops grow best together, develop crop varieties best suited to this form of farming, test how to boost populations of useful insects like ladybirds or earwigs, and work out how best to ensure that the organic matter content of soils slowly grows over time, rather than declining. They could even test whether biodynamic preparations actually work, or whether there is anything in the biodynamic practice of sowing according to the phases of the moon (always keep an open mind!).

Aside from the environmental benefits, a move towards more sustainable farming systems has the potential hugely to benefit human health directly. I have already mentioned how overconsumption of unhealthy and heavily processed food, comprising mainly grain, meat, sugar and oils, is impacting on our health, longevity and prosperity. Health services around the world are burdened with the huge costs of chronic illnesses directly resulting from poor diets. There are also major concerns over the long-term effects of chronic exposure to mixtures of pesticides in our food. We would be much healthier, and our economy would be greatly improved, if we could persuade people to improve their diets, ideally moving to eating much more seasonal, organic fruit and veg, and treating meat as an occasional luxury. This in turn would generate more demand for farms generating such produce.

Such changes in diet might come about through a bottom-up, grass-roots movement, as we are seeing with the growth of veganism among young people worldwide. Consumers are perhaps the most powerful group of all, for it is their purchases that fund the entire food system; if we stopped buying grain-fed, indoor beef, or indoor mass-reared chicken, it would cease to exist. If we stopped buying grapes flown in from South Africa or Chile, supermarkets would stop selling them. If we bought organic, local, seasonal fruit and veg then organic horticulture operations would proliferate around our cities. Given the major economic benefits of a healthier population, it would surely also be worthwhile for governments to take more action to promote healthy eating, such as ensuring that children are taught the benefits of a healthy diet from an early age (learning what a cauliflower is, for instance, despite its removal from the *Oxford Junior Dictionary*), and perhaps also by investing in public health campaigns. In the 1980s we had government campaigns warning of the dangers of HIV and AIDS, which some might argue was a relatively trivial threat to our health when compared to the harm done by poor diets. Governments could also consider more taxes on particularly unhealthy foodstuffs, like the existing 'sugar tax' currently levied on soft drinks in the UK. One could reasonably propose extending this to cover any highly processed food of little nutritional value, although that might end up including most of the food on sale in your local supermarket.

I have already mentioned how changes to subsidies and taxes might help steer us towards more sustainable food production, and also the need for R&D and an independent advisory service to support farmers. It would also be valuable if our governments would offer farmers free training to update themselves on the latest knowledge and research. Many professions have obligatory continuing professional development schemes, but there seems to be little available for farmers. In my experience farmers are much more likely to listen to other farmers than to anyone else, and so a network of demonstration farms which they can visit to exchange ideas and see different practices in action would be invaluable.

Of course, any move to change the way we grow food needs the buy-in of farmers themselves. Bringing them on board is clearly vital, but doing so can be tricky. For farmers, farming is more than just a job – it is a way of life – and a major part of their identity in a way that is quite different from most other professions. Predictably, any suggestion that farming may be in part responsible for insect declines, soil erosion or pollution of streams often provokes a defensive reaction and a digging in of heels. The National Farmers' Union in the UK in particular seems to be in complete denial, fighting hard against regulation and restriction of pesticides, and disputing the clear evidence that farmland wildlife is in a state of particularly rapid decline. It is a shame that environmentalists and farmers so often find themselves at loggerheads, for we all have the same common interests. It is not the fault of farmers that we have ended up with a global food production system that is vastly inefficient and damaging to our health and that of the environment. We could blame government policies and subsidies, supermarkets, stock market traders, the agrochemical industry, or the choices we make when we shop. We are all at fault, collectively. We all need farmers and farming, for without them we would starve. We all have common interests in ensuring that farmers can make a decent living and produce sufficient food, while practising methods that look after the soil, reduce carbon emissions and encourage healthy populations of pollinators. No farmer wants to hand over their farm to their children in a degraded, depleted state. It is in all our interests to recognise the problems and, together, to find a way to remedy them.

Trap-Building Ants

Deep in the Amazonian rainforests lives a small ant, *Allomerus decemarticulatus,* with a most unusual means of catching its prey. The ant is arboreal, nesting not underground but within special leaf pouches produced by the tree *Hirtella physophora,* some of the leaves of which curl up to form hollow chambers. The tree also provides the ants with sweet nectar which exudes from small swollen glands at the base of these leaves. This ant species is unique, so far as is known, in building traps to catch insect prey. It cuts hairs from the plant and weaves them together with purpose-grown fungal strands and regurgitated sticky secretions to produce a sponge-like structure built around and completely encasing the stems of the plant. The structure is riddled with tiny holes within which hundreds of the ants can hide, with just their heads protruding, and their sharp jaws open wide. If any large insect, such as a grasshopper or butterfly, is unfortunate enough to land on or walk over the structure, its feet and any other extremity within reach are immediately seized by the ants, which then quickly work to stretch the insect out, as if it were on a rack. Once it is completely immobilised, ants pour out from the spongy structure, carefully dismember their prey, and then carry the pieces back to their leaf pouch homes. The plant presumably benefits from being kept free of herbivorous insects.

Nature Everywhere

We Brits think of ourselves as a nation of nature lovers, with a strong emotional tie to our 'green and pleasant' land, regardless of whether we live in a thatched cottage in the countryside or in a flat in the city. We have a long tradition of both amateur and professional fascination with the natural world, stretching back to the likes of James Hutton, Gilbert White and Joseph Banks in the seventeen hundreds. Today, British professional ecologists include some of the best in the world, while an army of expert amateur enthusiasts collect data on our wildlife through a diversity of recording schemes: counting butterflies and bees, surveying ponds, ringing birds and so on. The BBC Natural History Unit in Bristol has produced some of the most inspiring and beautiful nature documentary series, with David Attenborough becoming a global champion, highlighting the plight of the natural world. The Royal Society for the Protection of Birds (RSPB) has over a million members, with the Wildlife Trusts not far behind at about 800,000, and many smaller but thriving charities focus on diverse wildlife groups from bumblebees to butterflies and mammals to plants.

Our collective enthusiasm for nature has led to the creation of numerous types of protected areas. We have 224 National Nature Reserves in the UK, together covering 94,000 hectares. We have reserves protected by international laws, such as Ramsar Convention sites (part of a global network of wetlands protected by an intergovernmental treaty) and Natura 2000 Sites (areas protected by European legislation). We also have Sites of Special Scientific Interest (SSSIs or triple-SIs), Special Areas of Conservation, and many local nature reserves, sometimes managed by the RSPB, one of the Wildlife Trusts, or other charities such as the Woodlands Trust. If that weren't enough we have Areas of Outstanding Natural Beauty, a network of National Parks, and the National Trust, which alone manages 250,000 hectares of land. In total, about 35 per cent of the land area of the UK has some kind of protected status.

It would be easy to conclude from these figures that our nature is in safe hands, and that we have done enough. Yet as we have seen our wildlife is in precipitous decline. A recent academic study, a major international collaborative project led by Professor Andy Purvis from London's Natural History Museum, analysed patterns of change in abundance and diversity of 39,000 species of plant and animal across 18,600 sites, globally, and calculated a 'Biodiversity Intactness Index' for each country. It estimated that the UK ranked 189th of 218 countries included in the study, making us one of the most nature-depleted countries in the world.

What has gone wrong? A major part of the problem is that most of the protections mentioned above are little more than an illusion. Much of the land in our National Parks and owned by the National Trusts is intensively managed farmland, with the usual blizzard of associated pesticides; it is no different to the rest of the countryside. Additionally, the Department of the Environment, Farming and Rural Affairs (DEFRA) recently estimated that only 43 per cent of our SSSIs were in 'good condition', and most are scarcely ever visited to check that they are being looked after. The most protected sites, like National Nature Reserves, can be destroyed to make way for a bypass or a new railway development such as the HS2 high-speed train link from London to Birmingham, if government decides it is appropriate. Even those nature reserves that are well managed and have been protected from development are small islands, surrounded by hostile environments, and beset by forces such as climate change, invasive species and drifting, seeping pollution.

Similar issues affect the 211,000 km² protected by the USA's sixty-two National Parks. These are supposed to be wilderness areas unaffected by man's activities, yet many are affected by oil and gas drilling, or by invasive species, while quite a few allow hunting, and climate change is affecting them all. The Everglades National Park, for example, is being damaged by over-extraction of water to irrigate crops, by fertiliser and pesticide pollution, and by no fewer than 1,392 different invasive species, spanning everything from Burmese pythons to spreading stands of Australian tea trees.

It is clear that trying to set aside areas for nature has not been adequate as a strategy to prevent biodiversity loss – though nature reserves undoubtedly have value – and that we need to do much more. We do not have to continue headlong towards environmental Armageddon, but to halt this process requires us to recognise that our current strategies are not working, and that we cannot carry on as we have in the past. It is not too late to save our planet, but to do so we need to learn to live alongside nature, to value and cherish it, to respect all life as equal to our own, especially the small creatures. If the rest of life on Earth is to thrive, we need to invite it in to live in our cities and on our farmland, to find ways to grow nutritious food that work with nature rather than by driving it out, that harness the power of insects and their kin to control pests, pollinate crops, and keep the soil healthy. We need to reduce our planetary footprint by reducing food waste and overconsumption of meat and processed foods, so that more land can be set aside for nature. With a switch to a predominantly vegetable diet, supplemented by small amounts of sustainably-harvested fish and grass-fed meat, we could greatly reduce the area of land used for human food production, and leave far more space for nature.

If we could do all this, then we could indeed 'rewild' substantial areas of the globe. It was way back in 1967 that the aforementioned E. O. Wilson and his colleague Robert MacArthur wrote *The Theory of Island Biogeography*, a book with a less-than-gripping title but which explained, for the first time, why small, isolated islands of habitat support few species, whereas large, connected islands support many more. Fifty years later, in 2016, E. O. Wilson followed this up by proposing, in his more readable book *Half-Earth*, that we should set aside half of the surface of the planet for nature. This may sound preposterous in an already crowded world, with a human population heading towards ten billion or more, until you consider that we currently grow three times the calories needed to feed everyone. It would certainly be possible to take vast tracts of land that are currently farmed out of production and still have easily enough to feed us all, particularly if we took out the least productive.

To take a simple example, let's look at cattle ranching in Brazil. Ranchers there are responsible for about 80 per cent of deforestation in Amazonia, releasing 340 million tons of carbon dioxide into the atmosphere each year, plus another 250 million tons of carbon dioxide equivalent in the form of methane from the livestock. Fires are used to clear the forest in the dry season, leaving the soil unprotected when the rains arrive so that much of it washes into rivers or blows away. Brazil now has an estimated 190 million cows, and exports beef all over the world, particularly to USA, Europe, and the growing Asian market. Globally, beef provides just 2 per cent of the calories we consume, yet 60 per cent of the world's agricultural land is used for beef production. Some of the Amazonian land cleared for cattle is fertile enough to sell on after a year or two of grazing to soy bean farmers, who grow beans mainly destined for export to USA or Asia, where it is fed to more cattle or to pigs. Other land has such thin soil that it is of little use for anything after a couple of years, so the cattle ranchers move on to clear more forest. The whole system makes a negligible contribution to feeding the world, while having a huge negative impact on both the global climate and on biodiversity. We need to find ways to stop this kind of slash-and-burn agriculture as a matter of extreme urgency, properly protect all of what remains of the Amazon forests, and try to restore the damaged land.

Closer to home, there are many areas of the UK that are not amenable to productive farming, and where nature might make better use of the land than we can. The Knepp project in West Sussex provides a well-known example. Formerly a large, farmed estate covering 1,700 hectares of mixed arable and pasture, it was running at a loss despite farm subsidies, primarily because the heavy clay soil is difficult to work and not especially fertile, so that crop yields were low. The owners decided to 'rewild', releasing some grazing animals – cows, ponies, deer and also pigs – and letting nature take its course. Nearly twenty years on, the land abounds with life, including thriving populations of some of our rarest birds and insects, including nightingales, turtle doves, purple emperor butter-flies, and a fantastic diversity of dung beetles feeding on the pesticide-free dung provided by the larger animals. There are many other areas of the UK where farmers struggle to make ends meet despite subsidies, and that produce little food. The environmentalist George Monbiot has suggested that much of the uplands of Britain, a large proportion of which are currently heavily grazed by sheep or deer, or managed by rotational burning to encourage red grouse, might be better set aside for nature. The overgrazing in these areas leads to low biodiversity, compacts the soil so that rainwater runs off rapidly and causes flooding downstream, and of course the animals are producing methane.

Some upland areas of north-western Britain were once clothed in temperate rainforests, comprising twisted, lichen-covered oaks, a habitat that has been almost entirely lost. In the Highland glens of Scotland there were magnificent, moss-festooned Caledonian pine forests, where capercaillie, pine marten and wildcat once thrived, but this too was mostly cleared. Other upland areas, on impermeable rocks that prevented the drainage of water, have given rise to slowly-growing beds of rich black peat, laid down over thousands of years. Much of the peatlands remain, but most have been damaged by ill-considered attempts to drain them. All of these habitats can support great biodiversity, capture carbon, and reduce flooding downstream, as well as providing tourism opportunities. These wide societal benefits would seem very likely to outweigh the production of small amounts of meat and wool.

Unfortunately, to say this has become controversial. Farmers who have kept sheep on the uplands for generations feel they are being threatened with eviction from their land and the loss of their way of life. This is understandable, but we cannot justify continuing to do something just because we have done it for a long time, especially if it requires others to subsidise it, and it is harmful to the environment. In any case, no one is proposing to re-enact the Highland Clearances, when many Scottish tenant farmers were evicted from the land by greedy landowners in the late 1700s. Nobody is forcing people from their land, and there is room for compromise. As at Knepp, small numbers of grazing animals are often used on nature reserves as a management tool, where they can be beneficial to biodiversity. The distinction between rewilding, Knepp-style, and extensive livestock farming is a blurry one, for at Knepp the livestock are culled and sold for meat. Historically, it was the action of low-intensity grazing by relatively small numbers of sheep and cattle that helped to create the wonderfully rich flora of, for example, the UK's chalk downlands, the limestone grasslands of southern France, and the high-altitude meadows of the Alps. Stocking density is key: small numbers of grazers, or occasional heavy grazing followed by periods of rest, might be optimal in some places, their methane production being offset by the benefits they provide for wildlife and for soil health.

The dream of many proponents of rewilding is to go one step further than Knepp has been able to, and reintroduce long-lost animals like the beaver, and large predators such as lynx, wolves and bears. Beaver can do a fantastic job of creating wetland habitats and reducing flooding downstream, while predators could, theoretically, remove the need for any human interference via culling of livestock. In Britain we have become used to living in a country with no large predators, so the very mention of reintroducing wolves can be enough to cause apoplexy in some quarters, but it is not so ridiculous. After all, farmers and wolves live alongside one another without insuperable problems in almost every country in continental Europe. Wolves have even recently returned to Holland, one of Europe's most densely populated small countries, while neighbouring Germany is home to an estimated 1,300 wolves in 105 packs, living alongside the human population of 83 million.

The tourism benefits from, for example, being able to offer the chance to see free-roaming wolves in a rewilded corner of Scotland would be very likely to far outweigh the financial losses associated with any harm to livestock, which in any case could be compensated.

I have a vision, in which the nation's gardens teem with wildflowers, bees, birds, butterflies and organic vegetables, and in which urban areas are free of pesticides. Our roundabouts, road verges and city parks are planted with wildflowers and flowering trees, and swarm with insect life. Our cities, which are mostly on fertile soils where early people chose to settle, are surrounded by allotments and small, labour-intensive biodynamic and permaculture farms producing abundant fresh fruit and veg sold direct into the cities, the crops pollinated by bees, wasps and hoverflies, and guarded against pests by an army of ladybirds, earwigs, soldier beetles and lacewings. Further into the countryside, farms are mixed, with small numbers of livestock, many more trees than at present, minimal use of pesticides, and a stronger focus on sustainability and soil health than on maximising yield. Farmers are supported in their endeavours by independent research, demonstration farms, ongoing professional training, and a network of independent advisers. Many have gone entirely organic, and all treat pesticides as a last resort. On the poorer soils that have never produced much food, rewilding projects along the lines of Knepp support rich biodiversity and provide a place for city dwellers to visit and experience nature running wild. Interspersed among this, our most special places, our nature reserves and SSSIs, are seen as sacrosanct, places where nature is given priority in perpetuity over the human appetite for more roads or factories or housing estates. These areas are supported by sufficient funds from government for them to be properly looked after. Our rivers are rewilded, with canalised banks removed so that they can meander as they once did, and clouds of mayflies shimmer above the water on a summer's eve. Beavers are encouraged to dam and create new wetlands, boosting biodiversity and reducing flooding downstream. In the more remote uplands, substantial wilderness areas have been created, where native forests can regenerate, and lynx, wolves and bears roam free. Crucially, in my dream world, humans do not regard their own needs as having priority over all others.

This might all seem far-fetched, but what is the point of dreams if you cannot let your imagination fly free? None of this is impossible, or even difficult. We have to change. We have to learn to live in harmony with nature, seeing ourselves as part of it, not trying to rule and control it with an iron fist. Our survival depends upon it, as does that of the glorious pageant of life with which we share our planet.

The Mantisfly

Even among the insect world, replete as it is with odd-looking creatures, the chimera-like appearance of the mantisfly stands out as peculiar.

The front half of these creatures very closely resembles a mantis, with powerful raptorial front legs, and a triangular head with large eyes. This is a wonderful example of what scientists call 'convergent evolution', whereby two unrelated creatures evolve to resemble one another as a solution to a common problem – in this case, efficiently catching and sub-duing passing insect prey. However, the mantisfly's hind half appears to belong to another creature entirely, with two pairs of diaphanous wings and a chubby, soft abdomen, superficially resembling the rear end of a lacewing or caddisfly, for those familiar with such creatures. In a few species the rear end is very wasp-like, with yellow and black stripes. Their life cycle, as with so many insects, is also remarkable. The young larvae of mantisflies lie in wait for a wolf spider, latching on as it scurries by, and either clinging to it externally, or – most commonly – climbing inside its lungs. They sustain themselves by sucking on the spider's haemolymph (blood) using piercing mouth parts. If the spider spins an egg sac, the larva climbs inside, and then completes its development inside the egg sac by sucking out the contents of the eggs, one by one.

To my regret these bizarre creatures do not occur in the UK, but they can be found in southern Europe, throughout the tropics, and in much of North America.

21

Actions for Everyone

There is no doubt that insects are in decline and, given their vital importance to the functioning of healthy ecosystems, and the critical role they play in our own food supply, this should be a cause of deep concern to all of us. Their declines are a sign that the fragile web of life on our planet is beginning to tear apart. It is too late for the St Helena giant earwig, and for Franklin's bumblebee, but it is not too late for most of the life on our planet. To save it, we need to act, and act now. Just one or two of us trying to help won't cut the mustard: we need an army of people, at all levels of society. Since you've got this far, I hope you realise the importance of taking responsibility, of us all getting involved in a concerted effort to change our relationship with the small creatures that live all around us. Here I give practical advice as to the many actions that we all need to take: some of them very simple, others a little harder, but all of them eminently possible; a manifesto for a greener, better world.

The actions suggested below are as seen from a UK perspective, but the large majority are relevant elsewhere.

Encouraging environmental awareness

We need to engender a society that values the natural world, both for what it does for us and for its own sake. The obvious place to start is with our children.

Action for national government

- Provide in-service training to enable teachers to confidently teach natural history. At present, many teachers simply do not have the knowledge to do this. A residential training centre where teachers could get an intensive crash course in nature studies would be invaluable.
- Provide every school with safe access to designated green space, so that children have the opportunity to interact with nature. Provide a network of advice and support to make school grounds more wildlife-friendly.
- Build education about natural history into the primary school curriculum, with at least one outdoor lesson per week. Done well, this should be the most fun lesson of the week, one that all the children eagerly look forward to.
- Introduce a secondary school qualification in natural history (e.g. GCSE in the UK).
- Twin schools with nature-friendly farms and provide funding for visits, so that all children get to visit a farm at least once per year and learn how food is produced and the challenges involved in farming.

Actions for all

- Vote for the party with the strongest and most convincing environmental policies, in both local and national elections. In the UK, our first-past-the-post system might make it seem that voting for the Green Party is a waste, but if mainstream parties see more people voting Green they will take on board their policies.

- Write to your political representatives at regular intervals, urging them to support green initiatives. Many politicians have little awareness of environmental issues, but you might begin to educate them!
- Spread the word, using any means at your disposal. Social media has enormous power; use whatever platforms you prefer to share interesting stories, activities or campaigns about insects, and post about what you and others are doing to help. Encourage your friends and neighbours to make their garden more insect-friendly, and to consider some of the other actions mentioned here.

Greening our urban areas

Imagine green cities filled with trees, vegetable gardens, ponds and wildflowers squeezed into every available space, and all free from pesticides. We could transform our urban areas, and what better time to start than right now, in our gardens.

Actions for gardeners and allotment holders

- Grow flowers that are particularly rich in nectar and pollen to encourage pollinators such as bees, butterflies and hoverflies. Lots of advice is readily available, for example in my book *The Garden Jungle* or online (e.g. shorturl.at/coxP4). Garden centres usually label pollinator-friendly plants, but beware that their plants often contain insecticides. You might try out the Bumblebee Conservation Trust's BeeKind tool to see how bee-friendly your garden is (easily located online).
- Grow food-plants for butterflies and moths, such as lady's smock, bird's foot trefoil, ivy and nettles.
- Reduce your frequency of mowing – allow your lawn (or part of it) to flower. You may be surprised by how many different flowers are already living in your lawn.

- Go one step further and create your own miniature wild-flower meadow. Simply stop cutting a patch of your lawn apart from an annual cut in September, and see what happens. A mix of long grass and (usually) some flowers will appear, which you can enhance by planting in more meadow wildflowers.

- Try to re-imagine 'weeds' such as dandelion as 'wild-flowers', and save yourself an awful lot of time spent weeding by allowing them to grow. 'Weedy' plants like dandelions, ragwort, hogweed and herb Robert are great flowers for pollinators.

- When wooden fence panels rot and collapse, as they in-evitably do after a few years, replace them with a hedge of mixed, native plant species. This is permeable to wild-life such as hedgehogs, provides food for caterpillars and pollinators, captures carbon as it grows, and will never need replacing.

- Buy or make a bee hotel, a fun project that children can get involved in. Lots of advice can be found online (e.g. shorturl.at/hAKLQ). In brief, all you need to do is create horizontal dead-end holes of about 8-millimetre diameter, either by drilling in to a block of wood or by bundling bamboo canes. Some commercial designs have windows, enabling you to peek in and see what the bees are doing inside their nests.

- Dig a pond and watch how quickly it is colonised by dragonflies, whirligig beetles, newts and pond skaters. Even tiny ponds, made by recycling an old sink or other waterproof container, can support abundant life. Make sure it is easy for any animal that falls in to climb out.

- Create a 'hoverfly lagoon', a small aquatic habitat for hoverflies to breed in. For advice on how to do so see https://www.hoverflylagoons.co.uk/

- Grow your own healthy, zero-food-miles fruit and veg. Every lettuce or carrot you grow saves you money and removes all the environmental costs of that food being grown elsewhere, packaged and transported to your plate.

- Plant a fruit tree. These are available in dwarf sizes suitable for tiny gardens, the smallest of which can be grown in a large pot on a patio or roof terrace. Fruit trees provide blossom for pollinators and fresh fruit for you. There is a mouth-watering array to choose from: apples, pears, plum, quince, apricot, mulberry, peach, fig and more.

- Avoid pesticide use in your garden; they really aren't necessary. If you simply leave pests alone, usually a ladybird, hoverfly larva or lacewing will come along and eat them before long. If you have ornamental plants that are continually attacked by insect pests, you are probably trying to grow the wrong plant. Weeds can either be accepted as wildflowers, hand-weeded, or an old carpet or other impenetrable material can be used to smother them.

- Use companion planting to encourage pollination of vegetable crops and to attract natural enemies of crop pests. For example, French marigolds seem to help deter whitefly from tomatoes, and borage attracts pollinators to strawberries.

- Leave a 'wild' corner for nature, where you do nothing at all: your own tiny rewilding project.

- Provide a brash pile or log pile, leaving the wood to moulder down, sprouting fungi and supporting a myriad of tiny decomposers

- Build a compost heap and recycle kitchen scraps, generating your own fertile compost while providing a home for worms, woodlice, millipedes and more.

Actions for national government

- Prohibit use of pesticides in urban areas, following the example of France, and of many large cities elsewhere, e.g. Ghent, Portland, Toronto. France banned use of pesticides in green public spaces in 2017, and banned the sale of pesticides to anyone apart from registered farmers from the beginning of 2020. This means that there will no longer be any domestic use of pesticides, no shelves stocked with rows of pesticides in garden centres, DIY stores and supermarkets. If the entirety of France can do this, so can other countries. Pesticide Action Network (PAN) UK can provide detailed advice for local authorities on alternative means of controlling pavement weeds with, for example, hot foam. However, I would argue that we should allow weeds to grow in cracks in the pavement, and get over our obsessive tidiness.

- Ban the use of neonicotinoid insecticides or fipronil (both highly potent insecticides) in flea treatments for pets and in ant baits. Both chemicals are regularly turning up in water samples from streams and rivers as a result of their use on pets. Fleas on pets can usually be kept under control by regularly washing of the pet's bedding, where the larval stages of the flea reside. If this fails there is a non-toxic silicone-based treatment called dimethicone that has been found to be effective.

- Introduce new legislation to ensure that nature-rich development becomes standard, providing real, measurable gains for wildlife, and to ensure that all new developments make a demonstrable, positive contribution to nature's recovery. These development plans should encompass the provision of habitat and improved connectivity of habitat for wildlife, and accessible green space, including space for community use such as allotments, taking into account effective water management, pollution and climate control.

- New flat-roofed buildings to include green roofs planted with pollinator-friendly plants. Some research is needed to identify suitable drought-tolerant, insect-friendly plants.
- Introduce legislation to ensure that all new golf courses maximise their potential to support biodiversity, including planting flowering, native trees and creating flower-rich meadow areas.
- Encourage the take-up of vegetable growing in home gardens and allotments through a public information campaign highlighting the health, environmental and economic benefits. This could be further supported by providing free training and vegetable seeds for novice vegetable growers, funded by diverting a tiny proportion of the current agricultural subsidy.
- Take steps to reducing light pollution. Most cities are lit up like Christmas trees at night, with some office blocks and roads lit throughout the night for little clear purpose. Motion-sensitive lighting could be used to ensure that both indoor and outdoor lights are turned off when nobody is near. Shielding can be used so that street lights or stadium lights only illuminate their target, preventing light from spilling out elsewhere. Funding for research to investigate whether certain frequencies of light are less disruptive to wildlife would also be valuable.

Actions for local government

- Prohibit the use of pesticides in urban areas if the national government has not done so (see above).
- Create wildlife areas in parks: meadows, ponds, plantings for pollinators, bee hotels and so on (see section on 'Actions for gardeners').

- Plant streets and parks with flowering, native trees, such as lime, chestnut, rowan, wayfaring tree and hawthorn.
- Plant fruit trees in urban green spaces like parks, providing blossom for pollinators and food for people.
- Reduce mowing of road verges and roundabouts, to allow wildflowers to flower, and removing the grass when it is cut (otherwise it may smother plants). Where possible, sow areas with an appropriate wildflower seed mix. All new road verges should be automatically sown with wildflower mixes.
- Purchase and/or dedicate land for allotments on city fringes and, when suitable areas become available, within cities. Recent evidence shows that allotments are the best areas in cities for pollinator diversity, while simultaneously providing healthy, zero-food-miles, no-packaging fruit and veg, and boosting the health of allotmenters (win, win, win).

Actions for all

- Write to the chairman of your local council, targeting issues of local relevance, such as pesticide use in your local parks and pavements, management of road verges for flowers, or creation of meadow areas in local green space.
- Get involved in, or set up, a local group such as the previously-mentioned On the Verge, to sow wildflower seeds and create flower-rich habitat in any unused or spare corners of land in urban areas, such as road verges and roundabouts.

Transforming our food system

Growing and transporting food so that we all have something to eat is the most fundamental of human activities. The way we do it has profound impacts on our own welfare, and on the environment, so it is surely worth investing in getting it right. There is an urgent need to overhaul the current system, which is failing us in multiple ways. We could have a vibrant farming sector, employing many more people, and focused on sustainable production of healthy food, looking after soil health and supporting biodiversity.

Actions for national government

- Redirect farm subsidies, the majority of which (about $11.5 billion per year in the USA) are currently given out as area-based payments so that the biggest farms get the bulk of the subsidy. These subsidies could instead be used to support production of the most nutritious food types (e.g. fruit and veg), and would only be given to farms using truly sustainable methods, and which set aside a minimum of 10 per cent of their land for nature. Smaller farms would get disproportionately more subsidy per unit area, making it easier for smaller operations to survive. Organic farms (including biodynamic and permaculture) would get a significant bonus. Payments to any one farm would be capped.
- Clearly define integrated pest management (IPM), as an approach to pest management that seeks to minimise pesticide use, treating it as a last resort, and then introduce legislation to make IPM compulsory (as it is already in the EU, albeit not enforced).
- Set targets for major reductions in weight of pesticides and fertilisers used, and number of applications per crop. Recent studies from France suggest that much pesticide use is unnecessary or at best insurance against unlikely events, and that farmers need independent advice and support to help them identify which pesticides they could do without.
- Introduce pesticide and fertiliser taxes (already used in Norway and Denmark), on the basis that polluters

should pay the full costs of their actions. The Danish system is based on taxes being proportionate to the environmental harm each chemical does, and could provide a useful model as most of the pesticides used in other countries are the same. Fertiliser use could also be discouraged by providing financial incentives for use of legume ley crops in rotations.

- Use revenue from the pesticide tax (above) to fund an independent advisory service to help farmers reduce pesticide use, develop an appropriate IPM system for their farm, or switch to a more sustainable farming system such as organic.

- Make all pesticide use open and transparent by obliging all farmers to submit the records they are already required to keep to a central open-access database. This would facilitate research into the impacts of pesticides on environmental and human health.

- Fund research and development into more sustainable farming methods – e.g. agroforestry, permaculture, organic, biodynamic – which currently receive minimal investment, but have the potential to be highly productive while supporting abundant biodiversity.

- Provide a system for training and support of continuing professional development for farmers, so that they have opportunity to update their skills and learn new techniques. This would include supporting peer-to-peer learning, since many farmers are actively investigating more sustainable farming methods and would benefit from mechanisms to help them share their knowledge.

- Set a target to have at least 20 per cent of land under organic farming by 2025 (Austria already has 23 per cent organic), and provide adequate financial support for farmers to make the transition.

- Remove support for biofuel crops. The scientific evidence suggests that there are far better ways of providing sustainable energy than via intensively farmed bio-fuel crops.

- Financially support large-scale rewilding projects on marginal lands which contribute little to food production – e.g. most uplands, and areas with infertile soils in the lowlands.
- Impose a tax on food that is flown in to the country, with the funds being used to support sustainable farming.
- Fund a public information campaign on the environmental and health benefits of eating locally grown, seasonal, fresh produce, and of reducing meat consumption.

Actions for local government

- Facilitate and support local food networks and farmers' markets, making it easier for farmers to sell their produce direct to the public.

Actions for farmers

- Recognise that there is a problem, and engage positively with government initiatives, conservation organisations and consumers in working to solve it. Like it or not, farming, along with most other human endeavours, is going to have to change rapidly in the twenty-first century. Business as usual is not an option, and neither is continuing 'traditional' farming approaches solely because they are traditional. Farmers will need to be willing to adapt fast, to consider and test alternative approaches to growing food: organic, permaculture, agroforestry systems and so on. They will also need to be prepared to engage with continuing professional development opportunities, and peer-to-peer learning, to ensure the effective spread of the latest knowledge and ideas. In marginal areas where farming is barely profitable even with subsidies, rewilding is an option to consider, and may provide a more reliable income.

Actions for all

- Recognise that every purchase we make has consequences. If we buy factory-farmed meat, we are supporting practices that are harmful to the environment, and which often involve animals living short, uncomfortable and sometimes downright awful lives. If we buy produce flown in from abroad, we are paying for the associated carbon emissions. Every piece of food packaging requires energy and resources to produce and dispose of (even when it is recyclable). Shopping has become an ethical minefield, but there are some simple principles to bear in mind.
- Support local, sustainable producers. Buy organic food, buy from local farmers' markets, or get an organic veggie box delivered. It is often argued that such food is unaffordable for many, but in the UK we spend just 10.5 per cent of income on food, down from about 50 per cent a hundred years ago. Veggie boxes can be surprisingly economical if you factor in the cost you would otherwise have to bear of driving to the supermarket, even more so if you cost in your own time.
- Buy seasonal produce.
- Buy loose fruit and veg.
- Be prepared to buy 'wonky' or imperfect fruit and veg.
- Avoid produce grown in a heated glasshouse, even when locally grown, as this can have an even larger footprint than food flown in from abroad.
- Cut down your meat intake, seeing it as a treat rather than an everyday part of your diet. Remember that chickens convert vegetable protein into animal protein much more efficiently than cows, pigs or sheep, and have lower associated greenhouse gas emissions. If buying red

meat, only buy grass-fed, outdoor-reared (it will usually say on the packet).

- Don't waste food: don't buy more than you need, or serve up portions that are too large; save and eat leftovers; use common sense rather than use-by dates to decide when food has gone off.

Improving protection of rare insects and habitats

Actions for national government

- Improve legal protection for insects. In the USA, the Endangered Species Act lists species that are threatened or endangered, and gives them additional legal protections. Despite comprising a significant majority of all known species on Earth, insects make up only 3 per cent of the species on this list, and most of these are butterflies. This is not because insects are any less threatened than other creatures, it is just that they are largely ignored. Most insects have no legal protection at present. Rare insects should be accorded equal weight to rare birds or mammals. Just because they are small does not make them unimportant.
- Properly fund government organisations responsible for wildlife conservation, such as the US Fish and Wildlife Service (FWS). The FWS is currently facing big budget cuts, including a halving of funding for management of endangered species.
- All of our remaining designated nature-rich areas should be regarded as sacrosanct. In the USA, approximately 14 per cent of the land area is designated for protection, including National Wilderness Areas, National Parks, National Forests and numerous state parks, but many have inadequate protection, and are exploited for hunting, forestry, grazing, mining and oil extraction. If government can give priority to short-term economic gain and over-

rule their protection, then eventually there will be none left. Mitigation measures, such as planting more trees elsewhere, are often used as an excuse to allow developments to go ahead, but it is self-evident that one cannot simply recreate rare habitat like ancient woodland.

- Provide adequate funding to support monitoring schemes so that we can develop accurate information as to which insects are most at threat, and where. An integral part of such funding would be to support the training of taxonomists – scientists with expertise in identification of insects. Taxonomy has been in decline for decades, meaning that there is now a severe shortage of experts able to identify many of our insect species.

- Provide funding for research into understanding the causes of insect declines; there is much that we do not understand, particularly with regard to the complicated interactions between different stressors that harm insects.

- Play a leading role in international initiatives to tackle climate change and biodiversity loss, setting an example of best practice for others to follow. In particular, we need a global initiative to prevent further deforestation in the tropics. It is often said, quite reasonably, that it is hypocritical for us rich Westerners living in countries long denuded of most of their wildlife-rich habitats to lecture poorer countries about caring for their environment. Yet much of the destruction is carried out by large multinational corporations, not by poor people trying to make a living. Whoever is doing it, we need to collectively work out how to stop it, and it will almost certainly require richer countries to be prepared to bear much of the cost.

Actions for all

- Join your local conservation organisation, or one of the many national conservation charities devoted to conservation. Your money will help to support their work. If you have time, get actively involved, for example by supporting their causes or joining their volunteer network.

- Become a wildlife recorder, joining a butterfly or pollinator monitoring program if there is one near you. You will be helping to collect valuable data on the changing populations of our insects, so helping to inform conservation strategies.

Acknowledgements

I would like to thank the many PhD students and postdocs with whom I have worked over the years; together we have managed to uncover just a few of the fascinating details of the secret lives of insects. I also thank my agent, Patrick Walsh, who saw something worth publishing in the first draft of *A Sting in the Tale*, and who eventually persuaded me to write *Silent Earth*. Finally, and above all, I thank my parents, who allowed and encouraged the eight-year-old me to fill our house with jam jars containing woolly bears, millipedes, earwigs, woodlice, crickets, and myriad other little creatures.

Further Reading

If you would like to know more about the subjects discussed in each chapter, there follows a selection of further reading. I have tried to include key scientific articles that provide the evidence underpinning our current understanding of insect declines and what we might do to reverse them. Sadly, many of these are not written for the layperson, and some of the technical jargon can be hard to follow. Nonetheless, non-specialists can usually glean the gist of an article without too much difficulty. Some of the articles are hidden behind paywalls, but if you are keen you can access most of them via the website Researchgate, which enables you to contact the authors directly and ask for copies of their work.

Chapter 1: A Brief History of Insects

Gould, S. J., *Wonderful Life: Burgess Shale and the Nature of History* (Vintage, London, 2000)

Grimaldi, D. and Engel, M. S., *Evolution of the Insects* (Cambridge University Press, Cambridge, 2005)

Wilson, E. O., *The Diversity of Life* (Penguin Press, London, 2001)

Chapter 2: The Importance of Insects

Ehrlich, P. R. and Ehrlich, A., *Extinction: The Causes and Consequences of the Disappearance of Species* (Random House, New York, 1981)

Garratt, M. P. D. et al., 'Avoiding a bad apple: insect pollination enhances fruit quality and economic value', *Agriculture, Ecosystems and Environment* 184 (2014), pp. 34–40

Garibaldi, L. A. et al., 'Wild pollinators enhance fruit set of crops regardless of honey bee abundance', *Science* 339 (2013), pp. 1608–11

Kyrou, K. et al., 'A CRISPR-Cas9 gene drive targeting *doublesex* causes complete population suppression in caged *Anopheles gambiae* mosquitoes', *Nature Biotechnology* 36 (2018), pp. 1062–6

Lautenbach, S. et al., 'Spatial and temporal trends of global pollination benefit,' *PLoS ONE* (2012), 7:e35954

Losey, J. E. and Vaughan, M., 'The economic value of ecological services provided by insects', *Bioscience* 56 (2006), pp. 3113–23

Noriega, J. A. et al., 'Research trends in ecosystem services provided by insects', *Basic and Applied Ecology* 26 (2018), pp. 8–23

Ollerton, J., Winfree, R. and Tarrant, S., 'How many flowering plants are pollinated by animals?' *Oikos* 120 (2011), *pp.* 321–6

Chapter 3: The Wonder of Insects

Engel, M. S., *Innumerable Insects: The Story of the Most Diverse and Myriad Animals on Earth* (Sterling, New York, 2018)

Fowler, W. W., *Biologia Centrali-Americana;* or, Contributions to the knowledge of the fauna and flora of Mexico and Central America, *Porter*, Vol. 2 (1894), pp 25–56

Hölldobler, B. and Wilson, E. O., *Journey to the Ants* (Harvard University Press, Harvard, 1994)

Strawbridge, B., *Dancing with Bees: A Journey Back to Nature* (John Walters, London, 2019)

Sverdrup-Thygeson, A., *Extraordinary Insects: Weird. Wonderful. Indispensable. The Ones Who Run Our World* (HarperCollins, London, 2019)

McAlister, E., *The Secret Life of Flies* (Natural History Museum, London, 2018)

Chapter 4: Evidence for Insect Declines

Bar-On, Y. M., Phillips, R. and Milo, R., 'The biomass distribution on Earth', *Proceedings of the National Academy of Sciences* 115 (2018), pp. 6506–11

Butchart, S. H. M., Stattersfield, A. J. and Brooks, T. M., 'Going or gone: defining "Possibly Extinct" species to give a truer picture of recent extinctions', *Bulletin of the British Ornithological Club* 126A (2006), pp. 7–24

Cameron, S. A. et al., 'Patterns of widespread decline in North American bumble bees', *Proceedings of the National Academy of Sciences* 108 (2011), pp. 662–7

Casey, L. M. et al., 'Evidence for habitat and climatic specialisations driving the long-term distribution trends of UK and Irish bumble-bees', *Diversity and Distributions* 21 (2015), pp. 864–74

Forister, M. L., 'The race is not to the swift: Long-term data reveal pervasive declines in California's low-elevation fauna', *Ecology* 92 (2011), pp. 2222–35

Fox, R., 'The decline of moths in Great Britain: a review of possible causes', *Insect Conservation and Diversity* 6 (2012), pp. 5–19

Fox, R. et al., *The State of Britain's Larger Moths 2013* (Butterfly Conservation & Rothamsted Research, Wareham, Dorset, 2013)

Fox, R. et al., 'Long-term changes to the frequency of occurrence of British moths are consistent with opposing and synergistic effects of climate and land-use changes', *Journal of Applied Ecology* 51 (2014), pp. 949–57

Goulson, D., 'The insect apocalypse, and why it matters', *Current Biology* 29 (2019), R967–71

Goulson, D. et al., 'Combined stress from parasites, pesticides and lack of flowers drives bee declines', *Science* 347 (2015), p. 1435

Grooten, M. and Almond, R. E. A. (eds), *Living Planet Report – 2018: Aiming Higher*, WWF, Gland, Switzerland, 2018)

Hallmann, C. A. et al., 'More than 75 per cent decline over 27 years in total flying insect biomass in protected areas', *PLoS ONE* 12 (2017), e0185809

Hallmann, C. A. et al., 'Declining abundance of beetles, moths and caddisflies in the Netherlands', *Insect Conservation and Diversity* (2019), doi: 10.1111/icad.12377

Janzen D. and Hallwachs, W., 'Perspective: Where might be many tropical insects?' *Biological Conservation* 233 (2019), pp. 102–8

Joint Nature Conservation Committee (2018), http://jncc.defra.gov.uk/page-4236

Kolbert, E., *The Sixth Extinction: An Unnatural History* (Bloomsbury, London, 2015)

Lister, B. C. and Garcia, A., 'Climate-driven declines in arthropod abundance restructure a rainforest food web', *Proceedings of the National Academy of Sciences* 115 (2018), E10397–E10406

Michel, N. L. et al., 'Differences in spatial synchrony and interspecific concordance inform guild-level population trends for aerial insectivorous birds', *Ecography* 39 (2015), pp. 774–86

Nnoli, H. et al., 'Change in aquatic insect abundance: Evidence of climate and land-use change within the Pawmpawm River in Southern Ghana', *Cogent Environmental Science* (2019), doi: 10.1080/23311843.2019.1594511

Ollerton, J. et al., 'Extinctions of aculeate pollinators in Britain and the role of large-scale agricultural change', *Science* 346 (2014), pp. 1360–2

Powney, G. D. et al., 'Widespread losses of pollinating insects in Britain, *Nature Communications* 10 (2019), p. 1018

Sanchez-Bayo, F. and Wyckhuys, K. A. G., 'Worldwide decline of the entomofauna: A review of its drivers', *Biological Conservation* 232 (2019), pp. 8–27

Semmens, B. X. et al., 'Quasi-extinction risk and population targets for the Eastern, migratory population of monarch butterflies (*Danaus plexippus*)', *Scientific Reports* 6 (2016), p. 23265

Shortall, C. R. et al., 'Long-term changes in the abundance of flying insects', *Insect Conservation and Diversity* 2 (2009), pp. 251–60

Seibold, S. et al., 'Arthropod decline in grasslands and forests is associated with landscape-level drivers', *Nature* 574 (2019), pp. 671–4

Stanton, R. L., Morrissey, C. A. and Clark, R.G., 'Analysis of trends and agricultural drivers of farmland bird declines in North America: a review', *Agriculture, Ecosystems and Environment* 254 (2018), pp. 244–54

Stork, N. E. et al., 'New approaches narrow global species estimates for beetles, insects, and terrestrial arthropods,' *Proceedings of the National Academy of Sciences* 112 (2015), pp. 7519–23

Van Klink, R., Bowler, D. E., Gongalsky, K. B., Swengel, A. B., Gentile, A. and Chase, J. M., 'Meta-analysis reveals declines in terrestrial but increases in freshwater insect abundances', *Science* 368 (2020), pp. 417–20

Van Strien, A. J. et al., 'Over a century of data reveal more than 80 per cent decline in butterflies in the Netherlands', *Biological Conservation* 234 (2019), pp. 116–22

Van Swaay, C. A. M. et al., *The European Butterfly Indicator for Grassland Species 1990–2013*, Report VS2015.009 (De Vlinderstichting, Wageningen, 2015)

Wepprich, T. et al., 'Butterfly abundance declines over 20 years of systematic monitoring in Ohio, USA', *PLoS ONE* 14 (2019), e0216270

Woodward, I. D. et al., *BirdTrends 2018: Trends in Numbers, Breeding Success and Survival for UK Breeding Birds*, Research Report 708 (BTO, Thetford, 2018)

Xie, Z., Williams, P. H. and Tang, Y., 'The effect of grazing on bumble-bees in the high rangelands of the eastern Tibetan Plateau of Sichuan', *Journal of Insect Conservation* 12 (2008), pp. 695–703

Chapter 5: Shifting Baselines

McCarthy, M., *The Moth Snowstorm: Nature and Joy* (John Murray, London, 2015)

McClenachan, L., 'Documenting loss of large trophy fish from the Florida Keys with historical photographs', *Conservation Biology* 23 (2009), pp. 636–43

Papworth, S. K. et al., 'Evidence for shifting baseline syndrome in conservation', *Conservation Letters* 2 (2009), pp. 93–100

Pauly, D., 'Anecdotes and the shifting baseline syndrome of fisheries', *Trends in Ecology and Evolution* 10 (1995), p. 430

Chapter 6: Losing their Home

Barr, C. J., Gillespie, M. K. and Howard, D. C., *Hedgerow Survey 1993: Stock and Change Estimates of Hedgerow Lengths in England and Wales, 1990–1993* (Department of the Environment, 1994)

Ceballos, G. et al., 'Accelerating modern human-induced species losses: entering the sixth mass extinction', *Science Advances* 1 (2015), e1400253

Fuller, R. M., 'The changing extent and conservation interest of lowland grasslands in England and Wales: a review of grass-land surveys 1930–84', *Biological Conservation* 40 (1987), pp. 281–300

Giam, X., 'Global biodiversity loss from tropical deforestation', *Proceedings of the National Academy of Sciences* 114 (2017), pp. 5775–7

Quammen, D., *The Song of the Dodo* (Scribner, New York, 1997)

Ridding, L. E., Redhead, J. W. and Pywell, R. F., The fate of semi-natural grassland in England between 1960 and 2013: A test of national conservation policy', *Global Ecology and Conservation* 4 (2015), pp. 516–25

Rosa, I. M. D. et al., 'The environmental legacy of modern tropical deforestation', *Current Biology* 26 (2016), pp. 2161–6

Vijay, V. et al., 'The impacts of palm oil on recent deforestation and biodiversity loss', *PLoS ONE* 11 (2016), e0159668

Chapter 7: The Poisoned Land

Bernauer, O. M., Gaines-Day, H. R. and Steffan, S. A., 'Colonies of bumble bees (*Bombus impatiens*) produce fewer workers, less bee biomass, and have smaller mother queens following fungicide exposure', *Insects* 6 (2015), pp. 478–88

Dudley, N. et al., 'How should conservationists respond to pesticides as a driver of biodiversity loss in agroecosystems?' *Biological Conservation* 209 (2017), pp. 449–53

Goulson, D., 'An overview of the environmental risks posed by neonicotinoid insecticides', *Journal of Applied Ecology* 50 (2013), pp. 977–87

Goulson, D., Croombs, A. and Thompson, J., 'Rapid rise in toxic load for bees revealed by analysis of pesticide use in Great Britain', *PEERJ* 6 (2018), e5255

Hladik, M., Main, A. and Goulson, D., 'Environmental risks and challenges associated with neonicotinoid insecticides', *Environmental Science and Technology* 52 (2018), pp. 3329–35

McArt, S. H. et al., 'Landscape predictors of pathogen prevalence and range contractions in US bumblebees', *Proceedings of the Royal Society B* 284 (2017), 20172181

Millner, A. M. and Boyd, I. L., 'Towards pesticidovigilance', *Science* 357 (2017), pp. 1232–4

Mitchell, E. A. D. et al., 'A worldwide survey of neonicotinoids in honey', *Science* 358 (2017), pp. 109–11

Morrissey, C. et al., 'Neonicotinoid contamination of global surface waters and associated risk to aquatic invertebrates: A review', *Environment International* 74 (2015), pp. 291–303

Nicholls, E. et al., 'Monitoring neonicotinoid exposure for bees in rural and peri-urban areas of the UK during the transition from pre- to post-moratorium, *Environonmental Science and Technology* 52 (2018), pp. 9391–402

Perkins, R. et al., 'Potential role of veterinary flea products in widespread pesticide contamination of English rivers', *Science of the Total Environment* 755 (2021), p. 143560

Pezzoli, G. and Cereda, E., 'Exposure to pesticides or solvents and risks of Parkinson's disease', *Neurology* 80 (2013), p. 22

Pisa, L. et al., 'An update of the Worldwide Integrated Assessment (WIA) on systemic insecticides: Part 2: Impacts on organisms and ecosystems', *Environmental Science and Pollution Research* (2017), doi.org/10.1007/s11356-017-0341-3

Sutton, G., Bennett, J. and Bateman, M., 'Effects of ivermectin residues on dung invertebrate communities in a UK farmland habitat', *Insect Conservation and Diversity* 7 (2013), pp. 64–72

UNEP (United Nations Environment Programme), *Global Chemicals Outlook: Towards Sound Management of Chemicals* (UNEP, Geneva, 2013)

Wood, T. and Goulson, D., 'The Environmental risks of neonicotinoid pesticides: a review of the evidence post-2013', *Environmental Science and Pollution Research* 24 (2017), pp. 17285–325

Yamamuro, M. et al., 'Neonicotinoids disrupt aquatic food webs and decrease fishery yields', *Science* 366 (2019), pp. 620–3

Chapter 8: Weed Control

Albrecht, H., 'Changes in arable weed flora of Germany during the last five decades', 9th EWRS Symposium, 'Challenges for Weed Science in a Changing Europe', 1995, pp. 41–48

Balbuena, M. S. et al., 'Effects of sublethal doses of glyphosate on honeybee navigation', *Journal of Experimental Biology* 218 (2015), pp. 2799–805

Benbrook, C. M., 'Trends in glyphosate herbicide use in the United States and globally', *Environmental Sciences Europe* 28 (2016), p. 3

Benbrook, C. M., 'How did the US EPA and IARC reach diametrically opposed conclusions on the genotoxicity of glyphosate-based herbicides?' *Environmental Science Europe* 31 (2019), p. 2

Boyle, J. H., Dalgleish, H. J. and Puzey, J. R., 'Monarch butterfly and milkweed declines substantially predate the use of genetically modified crops', *Proceedings of the National Academy of Sciences* 116 (2019), pp. 3006–11

Gillam, H., https://usrtk.org/monsanto-roundup-trial-tacker/monsanto-executive-reveals-17-million-for-anti-iarc-pro-glyphosate-efforts/ (2019)

Humphreys, A. M. et al., 'Global dataset shows geography and life form predict modern plant extinction and rediscovery', *Nature Ecology and Evolution* 3 (2019), pp. 1043–7

Motta, E. V. S., Raymann, K. and Moran, N. A., 'Glyphosate perturbs the gut microbiota of honeybees', *Proceedings of the National Academy of Sciences* 115 (2018), pp. 10305–10

Portier, C. J. et al., 'Differences in the carcinogenic evaluation of glyphosate between the International Agency for Research on Cancer (IARC) and the European Food Safety Authority (EFSA)', *Journal of Epidemiology and Community Health* 70 (2015), pp. 741–5

Schinasi, L. and Leon, M. E., 'Non-Hodgkin lymphoma and occupational exposure to agricultural pesticide chemical groups and active ingredients: A systemic review and meta-analysis', *International*

Journal of Environmental Research and Public Health 11 (2014), pp. 4449–527

Zhang, L. et al., 'Exposure to glyphosate-based herbicides and risk for non-Hodgkin lymphoma: a meta-analysis and supporting evidence', *Mutation Research* 781 (2019), pp. 186–206

Chapter 9: The Green Desert

Carvalheiro, L. G. et al., 'Soil eutrophication shaped the composition of pollinator assemblages during the past century', *Ecography* (2019), doi.org/10.1111/ecog.04656

Campbell, S. A. and Vallano, D. M., 'Plant defences mediate interactions between herbivory and the direct foliar uptake of atmospheric reactive nitrogen', *Nature Communications* 9 (2018), p. 4743

Hanley, M. E. and Wilkins, J. P., 'On the verge? Preferential use of road-facing hedgerow margins by bumblebees in agro-ecosystems', *Journal of Insect Conservation* 19 (2015), pp. 67–74

Kleijn, D. and Snoeijing, G. I. J., 'Field boundary vegetation and the effects of agrochemical drift: botanical change caused by low levels of herbicide and fertiliser', *Journal of Applied Biology* 34 (1997), pp. 1413–25

Kurze, S., Heinken, T. and Fartmann, T., 'Nitrogen enrichment in host plants increases the mortality of common Lepidoptera species', *Oecologia* 188 (2018), pp. 1227–37

Zhou, X. et al., 'Estimation of methane emissions from the US ammonia fertiliser industry using a mobile sensing approach', *Elementa, Science of the Anthropocene* 7 (2019), p. 19

Chapter 10: Pandora's Box

Alger, S. A. et al., 'RNA virus spillover from managed honeybees (*Apis mellifera*) to wild bumblebees (*Bombus* spp.)', *PLoS ONE* 14 (2019), e0217822

Darwin, C., *On the Origin of Species* (John Murray, London, 1859)

Fürst, M. A. et al., 'Disease associations between honeybees and bumblebees as a threat to wild pollinators', *Nature* 506 (2014), pp. 364–6

Goulson, D., 'Effects of introduced bees on native ecosystems', *Annual Review of Ecology and Systematics* 34 (2003), pp. 1–26

Goulson, D. and Sparrow, K. R., 'Evidence for competition between honeybees and bumblebees: effects on bumblebee worker size', *Journal of Insect Conservation* 13 (2009), pp. 177–81

Graystock, P., Goulson, D. and Hughes, W. O. H., 'Parasites in bloom: flowers aid dispersal and transmission of pollinator parasites within and between bee species', *Proceedings of the Royal Society B* 282 (2015), 20151371

Manley, R., Boots, M. and Wilfert, L., 'Emerging viral disease risks to pollinating insects: ecological, evolutionary and anthropogenic factors', *Journal of Applied Ecology* 52 (2015), pp. 331–40

Martin, S. J. et al., 'Global honeybee viral landscape altered by a parasitic mite', *Science* 336 (2012), pp. 1304–6

Chapter 11: The Coming Storm

Caminade, C. et al., 'Suitability of European climate for the Asian tiger mosquito *Aedes albopictus*: recent trends and future scenarios', *Journal of the Royal Society* Interface 9 (2012), pp. 2708–17

Kerr, J. T. et al., 'Climate change impacts on bumblebees converge across continents', *Science* 349 (2015), pp. 177–80

Lawrence, D. and Vandecar, K., 'Effects of tropical deforestation on climate and agriculture', *Nature Climate Change* 5 (2015), pp. 27–36

Loboda, S. et al., 'Declining diversity and abundance of High Arctic fly assemblages over two decades of rapid climate warming', *Ecography* 41 (2017), pp. 265–77

Pyke, G. H. et al., 'Effects of climate change on phenologies and distributions of bumblebees and the plants they visit', *Ecosphere* 7 (2016), e01267

Rochlin, I. et al., 'Climate change and range expansion of the Asian tiger mosquito (*Aedes albopictus*) in Northeastern USA: Implications for public health practitioners', *PLoS ONE* 8 (2013), e60874

Wallace-Wells, D., *The Uninhabitable Earth* (Penguin, London, 2019)

Warren, M. S. et al., 'Rapid responses of British butterflies to opposing forces of climate and habitat change', *Nature* 414 (2001), pp. 65–69

Wilson, R. J. et al., 'An elevational shift in butterfly species richness and composition accompanying recent climate change', *Global Change Biology* 13 (2007), pp. 1873–87

Chapter 12: Bauble Earth

Bennie, T. W. et al., 'Artificial light at night causes top-down and bottom-up trophic effects on invertebrate populations', *Journal of Applied Ecology* 55 (2018), pp. 2698–2706

Dacke, M. et al., 'Dung beetles use the Milky Way for orientation', *Current Biology* 23 (2013), pp. 298–300

Desouhant, E. et al., 'Mechanistic, ecological, and evolutionary consequences of artificial light at night for insects: review and prospective', *Entomologia Experimentalis et Applicata* 167 (2019), pp. 37–58

Fox, R., 'The decline of moths in Great Britain: a review of possible causes', *Insect Conservation and Diversity* 6 (2012), pp. 5–19

Gaston, K. J. et al., 'Impacts of artificial light at night on biological timings', *Annual Review of Ecology, Evolution and Systematics* 48 (2017), pp. 49–68

Grubisic, M. et al., 'Insect declines and agroecosystems: does light pollution matter?' *Annals of Applied Biology* 173 (2018), pp. 180–9

Owens, A. C. S. et al., 'Light pollution is a driver of insect declines', *Biological Conservation* 241 (2019), p. 108259

van Langevelde, F. et al., 'Declines in moth populations stress the need for conserving dark nights', *Global Change Biology* 24 (2018), pp. 925–32

Chapter 13: Invasions

Farnsworth, D. et al., 'Economic analysis of revenue losses and control costs associated with the spotted wing drosophila, *Drosophila suzukii* (Matsumura), in the California raspberry industry', *Pest Management Science* 73 (2016), pp. 1083–90

Goulson, D. and Rotheray, E. L., 'Population dynamics of the invasive weed *Lupinus arboreus* in Tasmania, and interactions with two non-native pollinators', *Weed Research* 52 (2012), pp. 535–542

Herms, D. A. and McCullough, D. G., 'Emerald ash borer invasion in North America: history, biology, ecology, impacts, and management', *Annual Review of Entomology* 59 (2014), pp. 13–30

Kenis, M., Nacambo, S. and Leuthardt, F. L. G., 'The box tree moth, *Cydalima perspectalis*, in Europe: horticultural pest or environmental disaster?' *Aliens: The Invasive Species Bulletin* 33 (2013), pp. 38–41

Litt, A. R. et al., 'Effects of invasive plants on arthropods', *Conservation Biology* 28 (2014), pp. 1532–49

Lowe, S. et al., *100 of the World's Worst Invasive Alien Species. A Selection from the Global Invasive Species Database* (IUCN Invasive Species Specialist Group, 2004)

Martin, S. J., *The Asian Hornet (Vespa velutina) – Threats, Biology and Expansion* (International Bee Research Association and Northern Bee Books, 2018)

Mitchell, R. J. et al., *The Potential Ecological Impacts of Ash Dieback in the UK* (JNCC Report 483, 2014)

Roy, H. E. et al., 'The harlequin ladybird, *Harmonia axyridis*: global perspectives on invasion history and ecology', *Biological Invasions* 18 (2016), pp. 997–1044

Suarez, A. V. and Case, T. J., 'Bottom-up effects on persistence of a specialist predator: ant invasions and horned lizards', *Ecological Applications* 12 (2002), pp. 291–8

Chapter 14: The Known and Unknown Unknowns

Balmori, A. and Hallberg, Ö., 'The urban decline of the house sparrow (*Passer domesticus*): a possible link with electromagnetic radiation', *Electromagnetic Biology and Medicine* 26 (2007), pp. 141–51

Exley, C., Rotheray, E. and Goulson D., 'Bumblebee pupae contain high levels of aluminium', *PLoS ONE* 10 (2015), e0127665

Jamieson, A. J. et al., 'Bioaccumulation of persistent organic pollutants in the deepest ocean fauna', *Nature Ecology & Evolution* 1 (2017), p. 0051

Leonard, R. J. et al. 'Petrol exhaust pollution impairs honeybee learning and memory', *Oikos* 128 (2019), pp. 264–73

Lusebrink, I. et al., 'The effects of diesel exhaust pollution on floral volatiles and the consequences for honeybee olfaction', *Journal of Chemical Ecology* 41 (2015), pp. 904–12

Malkemper, E. P. et al., 'The impacts of artificial Electromagnetic Radiation on wildlife (flora and fauna). Current knowledge overview: a background document to the web conference', A report of the EKLIPSE project (2018)

Shepherd, S. et al., 'Extremely low-frequency electromagnetic fields impair the cognitive and motor abilities of honeybees', *Scientific Reports* 8 (2018), p. 7932

Sutherland, W. J. et al., 'A 2018 horizon scan of emerging issues for global conservation and biological diversity', *Trends in Ecology and Evolution* 33 (2017), pp. 47–58

Whiteside, M. and Herndon, J. M., 'Previously unacknowledged potential factors in catastrophic bee and insect die-off arising from coal fly ash geoengineering', *Asian Journal of Biology* 6 (2018), pp. 1–13

Chapter 15: *Death by a Thousand Cuts*

Decker, L. E., de Roode, J. C. and Hunter, M. D., 'Elevated atmospheric concentrations of carbon dioxide reduce monarch tolerance and increase parasite virulence by altering the medicinal properties of milkweeds', *Ecology Letters* 21 (2018), pp. 1353–63

Di Prisco, G. et al., 'Neonicotinoid clothianidin adversely affects insect immunity and promotes replication of a viral pathogen in honeybees', *Proceedings of the National Academy of Sciences* 110 (2013), pp. 18466–71

Goulson, D. et al., 'Combined stress from parasites, pesticides and lack of flowers drives bee declines', *Science* 347 (2015), p. 1435

Potts, R. et al., 'The effect of dietary neonicotinoid pesticides on non-flight thermogenesis in worker bumblebees (*Bombus terrestris*)', *Journal of Insect Physiology* 104 (2018), pp. 33–39

Scheffer, M. et al., 'Quantifying resilience of humans and other animals', *Proceedings of the National Academy of Sciences* 47 (2018), pp. 11883–90

Tosi, S. et al., 'Effects of a neonicotinoid pesticide on thermoregulation of African honeybees (*Apis mellifera scutellata*)', *Journal of Insect Physiology* 93–94 (2016), pp. 56–63

Chapter 16: *A View from the Future*

Ghosh, A., *The Great Derangement: Climate Change and the Unthinkable* (University of Chicago Press, Chicago, 2017)

Lewis, S. and Maslin, M. A., *The Human Planet: How We Created the Anthropocene* (Pelican, London, 2018)

Ripple, W. J. et al., 'World scientists' warning to humanity: A second notice', *Bioscience* 67 (2017), pp. 1026–8

Wallace-Wells, D., *The Uninhabitable Earth*, op. cit.

Chapter 17: *Raising Awareness*

Booth, P. R. and Sinker, C. A., 'The teaching of ecology in schools', *Journal of Biological Education* 13 (1979), pp. 261–6

Gladwell, M., *The Tipping Point: How little things can make a big difference* (Back Bay Books, New York, 2002)

Morris, J. and Macfarlane, R., *The Lost Words* (Penguin, London, 2017)

Ripple, W. J. et al., 'World scientists' warning to humanity: A second notice', *Bioscience* 67 (2017), pp. 1026–8

Tilling, S., 'Ecological science fieldwork and secondary school biology in England: does a more secure future lie in Geography?' *The Curriculum Journal* 29 (2018), pp. 538–56

Chapter 18: Greening our Cities

Aerts, R., Honnay, O. and Van Nieuwenhuyse, A., 'Biodiversity and human health: mechanisms and evidence of the positive health effects of diversity in nature and green spaces', *British Medical Bulletin* 127 (2018), pp. 5–22

van den Berg, A. E. et al., 'Allotment gardening and health: a comparative survey among allotment gardeners and their neighbours without an allotment', *Environmental Health* 9 (2010), p. 74

Blackmore, L. M. and Goulson, D., 'Evaluating the effectiveness of wildflower seed mixes for boosting floral diversity and bumblebee and hoverfly abundance in urban areas', *Insect Conservation and Diversity* 7 (2014), pp. 480–4

Cox, D. T. C. and Gaston, K. J., 'Likeability of garden birds: importance of species knowledge and richness in connecting people to nature', *PLoS ONE* 10 (2015), e0141505

D'Abundo, M. L. and Carden, A. M., '"Growing Wellness": The possibility of promoting collective wellness through community garden education programs', *Community Development* 39 (2008), pp. 83–95

Goulson, D., *The Garden Jungle, or Gardening to Save the Planet* (Vintage, London, 2019)

Hillman, M., Adams, J. and Whitelegg, J., *One False Move: A Study of Children's Independent Mobility* (Policy Studies Institute, London, 1990)

Lentola, A. et al., 'Ornamental plants on sale to the public are a significant source of pesticide residues with implications for the health of pollinating insects', *Environmental Pollution* 228 (2017), pp. 297–304

Louv, R., *Last Child in the Woods; Saving Our Children from Nature Deficit Disorder* (Algonquin, Chapel Hill, NC, 2005)

Maas, J. et al., 'Morbidity is related to a green living environment', *Journal of Epidemiology and Community Health* 63 (2009), pp. 967–73

Monbiot, G., *Feral: Rewilding the Land, Sea and Human Life* (Penguin, London, 2014)

Moss, S., *Natural Childhood: A Report by the National Trust on Nature Deficit Disorder* (2012). Available online: http://www.lotc.org.uk/natural-childhood-a-report-by-the-national-trust-on-nature-deficit-disorder/

Mayer, F. S. et al., 'Why is nature beneficial?: The role of connectedness to nature', *Environment and Behavior* 41 (2009), pp. 607–43

Pretty, J., Hine, R. and Peacock, J., 'Green exercise: The benefits of activities in green places', *Biologist* 53 (2006), pp. 143–8

Rollings, R. and Goulson, D., 'Quantifying the attractiveness of garden flowers for pollinators', *Journal of Insect Conservation* 23: 803–17

Waliczek, T. M. et al. (2005), 'The influence of gardening activities on consumer perceptions of life satisfaction', *HortScience* 40 (2019), 1360–5

Warber, S. L. et al., 'Addressing "Nature-Deficit Disorder": A Mixed Methods Pilot Study of Young Adults Attending a Wilderness Camp', *Evidence-Based Complementary and Alternative Medicine* (2015), Article ID 651827

Wilson, E. O., *Biophilia* (Harvard University Press, Cambridge, MA, 1984)

Chapter 19: The Future of Farming

Badgley, C. E. et al., 'Organic agriculture and the global food supply', *Renewable Agriculture and Food Systems* 22 (2007), pp. 86–108

Baldock, K. C. R. et al., 'A systems approach reveals urban pollinator hotspots and conservation opportunities', *Nature Ecology & Evolution* 3 (2019), pp. 363–73

van den Berg, A. E. et al., 'Allotment gardening and health: a comparative survey among allotment gardeners and their neighbours without an allotment', *Environmental Health* 9 (2010), p. 74

Edmondson, J. L. et al., 'Urban cultivation in allotments maintains soil qualities adversely affected by conventional agriculture', *Journal of Applied Ecology* 51 (2014), pp. 880–9

Gerber, P. J. et al., *Tackling Climate Change Through Livestock – A Global Assessment of Emissions and Mitigation Opportunities* (Food and Agriculture Organisation of the United Nations, Rome, 2013)

Goulson, D., *Brexit and Grow It Yourself (GIY): A Golden Opportunity for Sustainable Farming* (Food Research Collaboration Food Brexit Briefing (2019), https://foodresearch.org.uk/publications/grow-it-yourself-sustainable-farming/

Hole, D. G. et al., 'Does organic farming benefit biodiversity?' *Biological Conservation* 122 (2005), pp. 113–30

Lechenet, M. et al., 'Reducing pesticide use while preserving crop productivity and profitability on arable farms', *Nature Plants* 3 (2017), p. 17008

Nichols, R. N., Goulson, D. and Holland, J. M., 'The best wildflowers for wild bees', *Journal of Insect Conservation* 23 (2019), pp. 819–30

Public Health England, 'Health matters: obesity and the food environment' (2017), https://www.gov.uk/government/publications/health-matters-obesity-and-the-food-environment/health-matters-obesity-and-the-food-environment–2

Seufert, V., Ramankutty, N. and Foley, J. A., 'Comparing the yields of organic and conventional agriculture', *Nature* 485 (2012), pp. 229–32

Willett, W. et al., 'Food in the Anthropocene: the EAT-*Lancet* Commission on healthy diets from sustainable food systems', *The Lancet* 393 (2019), pp. 447–92

Chapter 20: Nature Everywhere

Herrero, M. et al., 'Biomass use, production, feed efficiencies, and greenhouse gas emissions from global livestock systems', *Proceedings of the National Academy of Sciences* 24 (2013), pp. 20888–93

Monbiot, G., *Feral, op. cit.*

Newbold, T. et al., 'Has land use pushed terrestrial biodiversity beyond the planetary boundary? A global assessment', *Science* 353 (2016), 288–91

Purvis, A. et al., 'Modelling and projecting the response of local terrestrial biodiversity worldwide to land use and related pressures: the PREDICTS project', *Advances in Ecological Research* 58 (2018), pp. 201–41

Tree, I., *Wilding: The Return of Nature to a British Farm* (Picador, London, 2019)

Wilson, E. O., *Half-Earth: Our Planet's Fight for Life* (Norton, New York, 2016)

Index

Figures in *italics* refer to illustrations and charts.

'acclimatisation societies' 179
acid rain 216
Advantage (flea treatment) 109–10, 111
Advocate (flea treatment) 109–10, 111, 114*n*
aeroplanes 192
    contrails/chemtrails 192–3
Africa 16, 21, 63, 147, 159, 160, 220, 243
    East Africa 86–7, 147
    North Africa 147
    South Africa 21, 28, 134, 183, 266*n*, 275
Agricultural Development Advisory Service
    (ADAS) 273
agrochemicals *see* fertilisers; herbicides;
    pesticides
agroforestry 267–8
ahuahutle 22
algae 142, 143
allotments 75, 244, 252, 263–5, 266–7, 271,
    274, 284, 289–91, 293, 294
aluminium 192, 193
Amazon rainforests 75–6, 160, 162, 220,
    281–2
aminomethylphosphonic acid (AMPA)
    129–30
amphibians 13, 21, 48, 68, 124
    frogs 4, 21, 248
    newts 113, 240, 248, 290
    toads 28, 148, 175, 180, 183, 248
Anomalocaris 9–10
ant baits 292
Antarctica 161, 231, 259
Anthropocene, the 47, 48
antibiotic-resistant bacteria 30
ants 2, 14, 16–17, 23, 29, 194, 240

Allomerus decemarticulatus 277
    Argentine ants 183 *and n*
    fire ants 183
    honeypot ants 33
    leafcutter ants 72
    red ants (*Myrmica sabuleti*) 135
aphids 2, 21, 26, 57, 58, 85, 116, 167, 213,
    246
apples 31, 211, 265
arachnids 10, *see also* mites; spiders
Arctic permafrost 220
Areas of Outstanding Natural Beauty 279
arthropods 9–10, 11, *11*, 12, 53, 186
ash dieback 185
Attenborough, David 242, 278
Australia 19, 20, 90, 152
    bushfires 219, 242
    cattle farmers 27–9, 32
    dung beetles 28, 29
    Great Barrier Reef 221
    'green prescriptions' 252–3
    honeybees 147, 148, 150–51
    honeypot ants 33
    introductions and invasive species 26, 28–9,
        148, 150–51, 179–80, 181, 183, 186, 187
Austria 77, 262, 269, 296
avalanches *161*
avermectins 87, *88*
Aztecs 85

bacterial diseases (of insects) 146
bagworms *see* moths
Balbuena, Maria Sol 125
Bangladesh 160, 220

Banks, Joseph 278
Bar-On, Yinon 48
bats 13, 21, 35, 175, 212–13, 228
Bayer (company) 92, 95–6, 97, 98, *106*, 132
BBC Natural History Unit 278
beavers 283–4, 285
'bee hotels' 249, 290
bees 14, 16, 33, 58, 78, 80, 92, 101–3,
  112–13 and n, 122, 126, 129, 191, 194,
  195, 200, 213
  Andean 'giant golden bumblebees' (*Bombus
    dahlbomii*) 60, 154
  Asian honeybees (*Apis cerana*) 151, 153
  Bombus mesomelas 164, 166
  buff-tailed bumblebees 60, 148–9, 154, 186
  bumblebees 2, 3, 37, 58, 59, 60, 71, 92–4,
    95 and n, 103–4, 109, 110–12, 148–50,
    151, 152, 163, 164, 165, 166, 192, 194,
    200, 245, 249
  Chalicodoma nigripes 148
  common carder bumblebees 164
  common eastern bumblebees (*Bombus
    impatiens*) 149
  Cullum's bumblebees 59
  early bumblebees 164
  Franklin's bumblebees 60, 287
  garden bumblebees 164
  great yellow bumblebees (*Bombus
    distinguendus*) 58–9, 79
  honeybees 29, 30 and n, 90–92, 94, 95,
    96–7, 98, 102n, 103, 104, 109, 115–16,
    116, 125, 146–8, 151–4, 186, 194, 196,
    202, 219
  leafcutter bees (*Megachile*) 2, 102n, 148, 249
  mason bees (*Osmia*) 95 and n, 102n, 148,
    249
  orchid bees 84
  Pithitis smaragdula 148
  red-tailed bumblebees 164
  robotic bees 37, 219
  scabious mining bees
  short-haired bumblebees 59
  shrill carder bumblebees 78–9
  tree bumblebees 164
  Trigona carbonaria 150
  wild bees 30n, 58, 59, *60*, 74, 95n, 108
  yellow-faced bees 102n
beetles 14, 15n, 16, 61, 80, 89, 109
  bombardier 43, 156
  burying/carrion 29
  carabid 220
  death-watch 2
  dung 27–8, 29, 32, 117, 176, 221
  emerald ash borer 185
  ground 21, 25, 27, 57–8, 213, 267
  rove 27
  small hive 146
  soldier 213, 284
  whirligig 248, 290
Belgium 77, 149, 247

Benbrook, Charles 128, 129, 130
benzoquinones 156
Biden, President Joe 171
Biobest (company) 149
biodiversity 34, 187, 248, 250, 253–4, 265,
  266, 283, 285, 293
  and farming 5, 75, 77, 78, 83, 198, 236, 258,
    259, 261, 263 and n, 283, 285, 296
  loss of 20, 36, 47, 75, 78, 172, 206, 216,
    218, 228, 229, 279, 280, 281–2, 300
biofuel crops 260, 296–7
'bioprospecting' 30
Biotechnology and Biological Sciences
  Research Council (BBSRC) 100
birds 12–13, 38, 48, 61, 68–9, 105–6, 108n,
  228, 237–8, 248, 253–4
  bank swallows 61
  barn swallows 61
  blue tits 246
  bullfinches 179
  chaffinches 179
  chimney swifts 61
  common nighthawks (nightjars) 61
  corncrakes 77, 79
  crows 27
  cuckoos 61, 62, 69, 123–4
  goldfinches 122
  grey partridges 61, 123
  herons 21
  hoopoes 27
  house martins 35
  house sparrows 108n, 196
  lapwings 69
  moa 180
  nightingales 61, 179, 282
  osprey 21
  passenger pigeons 47, 70
  pheasants 230
  raptors 89
  red-backed shrikes 61
  red grouse 282
  seabirds 136, 137
  skylarks 179, 261
  sparrowhawks 21
  spotted flycatchers 61, 62
  starlings 21, 179
  swallows 35
  tits 175, 246
  turtle doves 282
  white-crowned sparrows 108n
  wrens 175
Blackmore, Lorna 249
Blattodea 16
blights 112
'blue baby syndrome' 143
Bolsonaro, President Jair 172
Booth, P. R. 239
Borlaug, Norman 140–41
Bornemissza, Dr George 28
Borneo rainforests 162

Bosch, Carl 140
Botías, Cristina 100–1, 102, 103, 104, 110
Boussingault, Jean-Baptiste 140
Boyd, Ian 51, 52, 114
Brazil 147, 152, 153, 162, 168, 281–2
Brexit 109, 229, 272
Brighton 248
Bristol 248
    University 264
British Trust for Ornithology 61
Buckland, William 137
'bug hotels' 248–9
Buglife (charity) 103*n*
bugs 8, 58, 237
    lantern bugs 38
    lightning bugs *see* fireflies
    marmorated stink bugs 181–2
    meadow spittlebugs 58
    shield bugs 38, 58, 181*n*
    stink bugs 21, 181*n*
bumblebees *see* bees
Burgess Shale fossils 9–10, *11*
butterflies 1, 2, 3, 14, 15, 16, 54, 55, 56, 65–6,
        66, 69, 71, 80, 109, 122, 129, 164, 165,
        248, 289, 299
    Adonis blue 81, 167
    birdwing 2
    brimstone 37
    chalkhill blue 79
    fritillaries 55, 77
    hairstreaks 55, 184
    large blue 36, 135
    large tortoiseshell 184
    large white 25–6
    meadow brown 55
    monarch (*Danaus plexippus*) 3, 13, 54–5,
        123, 203–5
    painted lady 13, 37
    peacock 55, 70, 225
    purple emperor 282
    silver-studded blue 166
    skipper 22, 122
    small blue 81
    small white 25–6
    wall (*Lasiommata megera*) 141–2
    white admiral 77
    white-letter hairstreak 184 *see also*
        caterpillars
Butterfly Conservation 164
Butterfly Monitoring Scheme 55
'buzz pollination' 149

cacao trees 35
Cactoblastis cactorum 26
caddisflies 21, 58, 105, 143, 207
California 54, 58, 130–32, 147, 162, 163,
        182, 183, 205, 219, 221, 252
    University of 131–2
Calyptra 38
Camponotus inflatus 33

Canada 9, 31, 54, 105, 157*n*, 168, 184, 192,
        195–6, 204, 247
cancers 89, 127–9, 131–2, 195–6, 236
CAP *see* Common Agricultural Policy
capillary action 100 *and n*
carbon dioxide emissions 80*n*, 143, 159–60,
        162 *and n*, 171, 217, 218, 219–20, 266
        and n, 271, 276, 281, 283, 290, 298
cardenolides 204
Caring for God's Acre (charity) 250*n*
Carson, Rachel: *Silent Spring* 3, 4, 65, 88, 89,
        90, 260
caterpillars 14, 16, 21, 26, 38, 80, 115, 123,
        142, 166–7, 245, 290
    bagworms 207
    centre-barred sallow moth 185
    cinnabar moth 1
    common blue butterfly 245
    elephant hawk moth 188
    emerald moth 188
    large blue butterfly 135
    lobster moth 188
    maguey worms 22
    monarch butterfly 173, 204
    mopanie worms 21, 23
    orange tip butterfly 245
    pine processionary moth 145
    silkworms 21–2
    skipper butterfly 122
    small copper butterfly 142
    small heath butterfly 142
    small white butterfly 256
    snailcase bagworm 207
    speckled wood butterfly 142
    swallowtail butterfly 38, 188
    tiger moth 124
    tobacco hornworm moth 142
cats 110, 111, 137–8
cattle farming 22, 23, 27–9, 32, 48, 141, 263,
        281, 282
cemeteries 250 *and n*
centipedes 10
Centre for Ecology & Hydrology (CEH),
        Oxfordshire 59, 95
cereals, breakfast 126
chalk downs 77, 78, 82, 283
chikungunya 168
children 5, 37, 49, 69–70, 210, 232, 246
    and diet 262, 275
    educating 5, 230, 236–41, 249, 287–8, 290
    and glyphosate 126–7
    and nature 251, 252, 255
Chile 137, 148, 149, 154, 265
China/the Chinese 22, 31, 77, 80*n*, 85, 133,
        137, 154, 155, 160, 219, 168, 174, 182
chocolate 23, 214, 220
cicadas, periodical 226
cities/towns 4, 5, 8, 19, 167–8, 257
    greening 244, 252, 289–91
    and light pollution 174–7, 293

pesticide-free 247–8, 291–2
and rising sea-levels 160–61, 219, 224
sparrows 196
wildflower patches 190, 249–50, 154 *see also* allotments; gardens/gardeners; parks
climate change 3, 31, 74, 77, 82, 123, 143, 144, 157–72, 192, 193, 194, 204–5, 212, 217, 218, 219, 220, 232, 242, 300
cockroaches 8, 13, 15, 16, 19, 67, 152
coffee-growing 168
Coleoptera 16
Collembola 26–7
Colombia 134, 168
Colorado, USA 163, 193
Common Agricultural Policy (CAP) 271–3
companion planting 291
compost 26, 117, 118, 139, 240, 247, 248, 254, 264, 266–7, 269, 291
Congo, the 61, 174, 177
Constable, John: *Salisbury Cathedral* ... 184
coppicing 77–8
coprolites 137
coral reefs 41, 70, 221, 242
Cotesia glomerata 26
COVID-19 22, 197, 219
cows *see* cattle farming
cowslip wine 69
crickets 21, 22
  bush crickets 37
  giant wetas 35–6, 183
  wart-biter crickets 36, 79
crops
  biofuel 260, 296–7
  cereal 24, 101, 126, 167, 171, 172, 220, 259–60, 264, 271
  and disease 85, 112, 181, 263
  flowering crops and neonicotinoids 94–5, 97, 100, 101, 104, 108, 109; *and see* oilseed rape, red clover, sunflowers *below*
  and fungicides 112
  genetically modified (GM) 121–2, 123, 197–8, 203–4
  hand-pollinating 31, 212
  oilseed rape 92–4, 95, 101, 104, 203, 264, 267, 271
  and pest control by insects 25–6
  pollination of 24–5, 30–31, 37
  red clover 149–50 *and n*
  'Roundup-ready' 121–2, 198
  sunflowers 90, 92, 94
Cruiser (insecticide) 114*n*
crustaceans 11, 12, 105, 107, 186, 190

Darwin, Charles 43, 84, 137, 150
Decker, Leslie 204
decomposers 26–7
deer 180, 213, 269, 282
deforestation 61, 75–7, 205, 206, 216, 217, 219, 244, 281–2, 300
Denmark 273, 295

Department of the Environment, Farming and Rural Affairs (DEFRA) 51, 100, 114, 115, *121*, 279
diabetes 214, 223, 252, 259, 262
diesel fumes 191, 203
diet, human 23, 24, 280
  healthy 215, 259, 275
  and insect consumption 21–3
  poor 91, 259, 262, 274 275
  vegan 263*n*, 275
Dijk, Tessa van 105
dinosaurs 12*n*, 137
Diptera 16
diseases
  crop and plants 85, 112, 181, 263
  human 8, 19, 22, 25*n*, 133, 168, 192, 206, 219, 224, 254, 274, *see also* cancers; malaria
  insect 60, 74, 91, 97, 112–13, 125, 146–7, 151–5, 170, 186, 199, 201, 202–3
  tree 181, 184–5
dragonflies 5, 13, 65, 143, 248, 290
Drosophila bifurcata 40
droughts 30, 82, 160, *161*, 162, 166–7, 197, 214, 220, 224, 293
dung beetles *see* beetles
dung flies *see* flies
Dutch elm disease 184–5

earwigs 2, 25, 117, 119, 152, 213, 246, 274, 284
  giant 35, 287
Easter Island 76, 183
'eco-anxiety' 231
ecology 239–40
'ecopsychology' 251
Egypt/Egyptians, ancient 85, 137–8 *and n*, 147, 148, 243
Ehrlich, Paul 31, 206
electromagnetic radiation 194–7, 203
electrostatic charges 200
EPA *see* United States: Environmental Protection Agency
Ethiopia 139, 168
European Commission 104
European Food Safety Authority (EFSA) 94, 104, 108, 127
European Parliament 94
Exley, Chris 192
Extinction Rebellion movement 230–31, 241
extinctions 10, 12*n*, 47, 205, 206, 217, 228, 230
  insect 4, 30, 35–6, 58, 59
  'megafauna' 47–8
  plant 32, 124–5

Fabre, Jean-Henri 145
Faldyn, Matt 204
famines 79, 212, 222

farming/farms 79–80, 83, 165, 221, 257–76, 297–8
   and biodiversity 5, 75, 77, 78, 83, 198, 236, 258, 259, 261, 263 *and n*, 283, 285, 296
   biodynamic 267, 269–71, 272, 295
   cattle *see* cattle farming
   and GM technology 197–8, 203–4, 219
   and the 'green revolution' 140–41
   and greenhouse gases 80, 143–4, 263, 272, 298
   and insect decline 36, 52–3, 55, 57, 66, 70, 75, 78, 79–80,
   of insects 22–3
   and 'integrated pest management' (IPM) *106*, 260–61, 271, 295
   organic 235, 262, 263, 272, 273, 284, 295, 296, 297, 298
   permaculture 267, 268–9, 270–71, 295
   and schools 240, 288
   and 'sharing-sparing debate' 83
   and soil erosion 80 *and n*, 214, 217, 221, 229, 266–7
   and subsidies 79, 236, 240, 258, 261, 265, 271–3, 275, 276, 282, 283, 293, 295, 298
   sustainable 4–5, 75, 259–60, 263*n*, 272, 273, 274, 280, 284, 295, 296, 297, 298
   and training of farmers 275–6, 284, 296
   upland 282–3 *see also* crops; fertilisers; herbicides; pesticides
Fereday, R. W. 150*n*
fertilisers 79, 80, 136–44, 206, 214, 258, 261, 268, 273, 295
   phosphates 136–9, 140, 214
fireflies 2, 18, 176
fires *see* wildfires
fish 4, 21, 48, 68, 71, 105, 206, 216, 218, 221, 222, 280
   eel 23–4, 107
   mosquito fish 183
   salmon 23
   smelt 23–4, 107
   trout 23
Flat Earth Society, UK 157*n*
flea treatments 109–10, 111, 292
flies 8, 14, 15, 16, 59, 61, 105, 117, 169, 175
   bluebottles 19, 29
   dung flies 27–8
   fever flies (*Dilophus febrilis*) 57
   flesh flies 29
   fruit flies (*Drosophila bifurcata*) 40
   fungus gnats 178
   greenbottles 29
   house flies 8, 14, 19, 167, 213
   midges 34–5 *and n*
   spotted wing drosophila 182
   stoneflies 143
   twisted wing flies 40 *see also* hoverflies

flooding/floods 82, 160, *161*, 162, 163, 166, 219–20, 222, 224, 242, 282
   reducing 268, 283, 284, 285
Florida, USA 160, 219, 220
flowers and plants *see* wildflowers and plants
food
   insects as 21–3
   processed 259, 263, 274
   producing 258–9, *see also* farming/farms
   reducing waste 4, 262, 263, 280
   transforming systems 294–9
food chains 20–21, 23–4, 89, 90, 123, 213, 228
forests *see* deforestation; rainforests, tropical; wildfires
fossil fuels 82, 143, 161, 162*n*, 171, 216, 263
fossils 9, 10, 12, 13, 137
foulbrood 146
Fowler, William Weekes *39*
foxes 179–80, 230*n*
France 77, 90–91, 92, 152, 155, 161, 182, 247–8, 271, 292, 295
Fray, Dennis 54
freshwater habitats 21, 23, 48*n*, 65, 105–7, 142–3, 217
froghoppers 58, 61, 80
frogs *see* amphibians
Frontline (flea treatment) 110, 111
fruit growing 5, 25, 31, 211, 212, 213, 258, 259, 263*n*, 264, 265–6, 267, 268, 270, 271, 274, 275, 284, 290–1, 294, 298
   and pests 182, 213
fungi 72, 291
   chalkbrood 146
   stonebrood 146
fungicides 111–13, 201–2, 203
   chlorothalonil 112–13
   ergosterol biosynthesis-inhibiting (EBI) fungicides 113
fungus gnats 178

Gambia 171
Gandalf (insecticide) 114*n*
gardens/gardeners 241, 244, 245–7, 248–9, 252, 253–4, 289–91, 293
geckos 175
gene drive technology 36 *and n*
genetically modified organisms (GMOs)/GM technology 36*n*, 121–2, 123, 197–8, 203–4, 219
geoengineering 192, 193–4, 218
Germany 3, 52–4, 61, 65, 77, 87, 91, 95, 124, 126–7, 132, 168–9, 217, 234–6, 242, 259, 271, 284 *see also* Krefeld Society
Glacier National Park, Montana 160
Gladwell, Malcolm: *The Tipping Point* 241
Glastonbury 248
glow-worms 176–7
   sticky cave (*Arachnocampa lunimosa*) 178
   see also fireflies
glyphosates *see* herbicides

golf courses 75, 77, 165, 250, 293
Gómez González, Homero 205
Göppel, Josef 235
grasses 24, 78, 141
    buffelgrass 186
    cock's foot grass 141
    couch grass 121, 122
    goose grass 212
    reed canarygrass 186
    rye grass 141
grasshoppers 2, 5, 13, 15–16, 21, 40, 43, 61,
        70, 77
    'inago' 22
grasslands 3, 52–3, 55, 78, 206, 237, 250, 283
Great Barrier Reef 221
'green prescriptions' 252–3
'green revolution' 140–41
greenhouse gases 23 *and n*, 80, 143–4, 157,
        158, 159–60, 162 *and n*, 171 *and n*, 216,
        217, 218, 219–20, 266 *and* n, 271, 272,
        276, 281, 283, 290, 298
Greenland 161, 162, 169, 259
groundwater 82, 98, 113, 216
guano 136–7

Haber, Fritz 140
habitat loss 74, 77, 83, 201 *see also*
    deforestation
Haldane, J. B. S. 41*n*
Hallucigenia 9
Hardeman, Edwin 131
Harvard University 193
Hawaii 152, 180–81, 186–7
heaths, lowland/heathland 79, 83, 166
heatwaves *161*, 166
heavy metal poisoning 189–90, 193, 218
hedgehogs 34, 70, 213, 225, 290
hedges and hedgerows 79, 80, 141, 258
Hemiptera 58
Henry, Mickaël 94, 96
Henslow, Reverend John Stevens 137
herbicides 112, 120, 125
    dicamba 123, 203
    glyphosate 114*n*, 120–22, *121*, 123,
        125–33, 198, 203, 236
    paraquat 133–4
Hernández Romero, Raúl 205
Herndon, Marvin 193
Hill, Octavia 251
Hirohito, Emperor 22
Holmgren, David 268
honeybees *see* bees
hornets, Asian 155, 182, 183
hoverflies 3, 5, 25, 59, *60*, 80, 155, 203, 213,
        220, 245, 246, 249, 267, 284, 290, 291
HS2 high speed train link 279
Hudson, Quebec 247
Humboldt, Alexander von 136–7
hurricanes 160, 219
Hussein, Saddam 189

Hutton, James 278
hydroquinones 156
Hymenoptera 16

I. G. Farben 87, 92
IARC *see* International Agency for Research
    on Cancer
imidacloprid *see* pesticides
Imperial Chemical Industries (ICI) 133
India 1, 80*n*, *130*, 134, 148, 174, 219
insecticides *see* pesticides
Institute of Terrestrial Ecology 55
'integrated pest management' (IPM) *106*,
        260–61, 271, *295*
International Agency for Research on Cancer
        (IARC) 127, 128, 129, 130, 132, 196
International Union for the Conservation of
        Nature (IUCN) 63–4
invasive species 26, 28, 147–8, 150–51,
        179–81, 183, 186–7, 280
IPM *see* 'integrated pest management'
Irwin, June 247
Italy 77, 98, 157*n*, 166, 271
IUCN *see* International Union for the
    Conservation of Nature

Jakarta, Indonesia 160, 219
Janzen, Dan 61, 63
Japan 22, 23, 38, 77, 119, 134, 148, 149, 152,
        154, 174, 182, 183, 184, 219, 247, 253
    Lake Shinji study 23, 106–7, *107*
Johnson, Dewayne 130–31, 132
Jonghe, Dr Roland De 149

Keele University, Staffordshire 192
Kenya 31, 168
Klink, Roel van 64–5
Knepp Wildland Project, West Sussex 282,
        283, 285
Krefeld Society 49–51
    study 51–2, 53, 57, 65, 81, 83, 168–9, 174,
        234–5
Krupke, Christian 97, 99
kudzu vines 186
Kunin, Professor William 53

lacewings 25, 155, 220, 246, 284, 291
ladybirds 25, 155, 213, 220, 246, 267, 274,
        284, 291
    harlequin 183, 203
lakes: pollution 23, 106–7, *107*, 77, 105, 118,
        142–3, 206
landslides/mudslides *161*, 163
larvae
    bagworm 207
    bee 97, 152, 191
    butterfly 135
    caddisfly 21, 58
    flea 292
    hoverfly 291

larvae (*cont.*)
  insect 14, 15, 35
  mantisfly 286
  sticky cave glow-worm 178
  twisted-wing fly 40
  wasp 256
latex 173
Lawes, Sir John Bennet 138, 139n
lawns 246, 289–90
Leanchoilia 9, 11
leeks 265–6
Leopold, Aldo 30, 35
Lepidoptera 16
Lewes 248
lice
  body 85
  book 40 *see also* woodlice
lichens 185
light pollution 174–7, 201, 293
Lister, Bradford C. 63
  Puerto Rico studies 63, 169–70
Litt, Andrea 186
lizards 13
  anolis 63
  horned 183
  sand 77
Loboda, Sarah 169
local government 293–4, 297
locusts 16, 85, 243
London 79, 219, 248
  Imperial College 68
  Zoological Society 48
Louisiana State University 204
Louv, Richard: *Last Child in the Woods* 251
Lucas, Caroline, MP 241
Lycaenidae 135

MacArthur, Robert *see* Wilson, E. O.
Macfarlane, Robert: *The Lost Words* 238
McGill University, Canada 169
maggots 14, 15, 16, 29
Malaise, René 49
  Malaise traps 49–50
malaria 86, 89, 168, 213, 224
Maldives, the 160, 220
mammals 21, 28, 47, 48, 64, 68, 118, 180,
    262, 268 *see specific species*
mantises 13, 21
mantisflies 286
Maori, the 180
Marianas Trench 190
Marriot, David 54
Marshall Islands 160
marsupials 28
  greater and lesser bilby 180
Mavrik (insecticide) 114n
mayflies 13, 61, 105, 143, 176, 285
meadows 77, 78, 79, 83, 141, 144, 248, 290
meat consumption 4, 259, 263 *and* n, 274–5,
    280, 297, 298–9

medical uses of insects 30
'megafauna extinctions' 47–8
Meganeura 13
mercury 189–90
metaldehyde 80, 142
metamorphosis 14–15
methane production 23 *and* n, 143, 159, 217
'Mexican caviar' 22, 23
Mexico 22, 54, 55, 90, 134, 149, 173,
    204, 205
Miami, Florida 219
microsporidia 147
midges 34–5 *and* n, 65
mildews 112
milk
  breast 89–90
  cow's 89–90
milkweeds 173, 204
millipedes 10, 11, 12, 291
minerals, soil 136
Mitchell, Edward 108
mites 10, 146, 155 *see also* Varroa
mobile phones 194–6
Mollison, Bill 268–9
Monbiot, George 282
  Feral 251
Monks Wood Research Station,
    Cambridgeshire 55
Monroe, Mia 54
Monsanto 128, 131, 132
Moon, Jennifer 131
Moon, the 175–6
Moor, Jennifer 131
Morocco 139, 171
Morrissey, Christy 105
mosquitoes 8, 34, 36 *and* n, 65, 86, 183,
    212–13
  Anopheles 36, 168
  Asian tiger (*Aedes albopictus*) 168
  yellow fever (*Aedes aegypti*) 167–8
moths 14, 15n, 16, 21, 38, 54, 55–7, 58, 69,
    80, 109, 175, 177, 289
  Atlas moths 1
  bagworms 207
  box tree moths 181
  centre-barred sallow moths 185–6
  cinnabar moths 1
  clothes moths 19
  emperor moths 21
  green moon moths 1
  hawk moths 192
  Hemiceratoides hieroglyphica 38
  peacock moths 1
  silk moths 29
  sloth moths 38
  tiger moths 124
  vampire moths 38
  wax moths 146 *see also* caterpillars
Motta, Erick 125
mudslides 163
Müller, Paul Hermann 86, 87

Mumbai, India 219
Munich: Technical University 52–3

National Farmers' Union 276
National Nature Reserves 278, 279
National Parks 279
National Trust 251, 279
Natura 2000 Sites 279
nature, contact with: and health benefits
    249–55
Nature (journal) 51
nature reserves 82, 97, 143, 165, 259, 278,
    279–80, 283, 285
    city 254, 257, 264
    German 49–50, *50*, 52, 81, 168, 174, 217,
    259
nematode worms 146
neonicotinoids *see* pesticides
Netherlands 57–8, 65–6, 66, 77, 105–6, 142,
    252, 264–5, 284
New York 219
New Zealand 2, 35–6, 80*n*, 147, 148, 149–50
    and n, 152, 178, 179, 180, 183, 186–7,
    252
newts *see* amphibians
Nicholls, Beth 100, 102, 103, 104, 110,
    264
Niño, El 162
nitrates 80, 139–40, 143, 144
nitrocellulose 140
nitrogen 136, 139–40, 142, 143, 150N
nitroglycerin 140
nitrous oxide 143–4
non-Hodgkin's lymphoma 127, 131, 132
Norway 273, 295
Nosema bombi 112
    N. ceranae 112, 153–4

obesity 214, 223, 259, 262
oceanic 'dead zones' 216, 217
On the Verge (conservationists) 249, 294
Opabinia 9, *11*
Ophryocystis elektroscirrha (parasite) 204
orchards *88*, 211, 258
Orthoptera 16
Osaka, Japan 219
Oxford Junior Dictionary 238, 275
ozone layer 144, 216

palm oil production 75, 76
parasites, insect 146, 147, 151, 155, 204
parasitoids 256
Paris Agreement (2016) 170–71, 172
parks, city 4, 19, 117, 118, 244, 247, 248,
    249, 250, 252, 254, 284, 293, 294
peat fires 162 *and n*, 219
peatlands 77, 162, 283
Pembrokeshire 78
Penn State University, USA 248
Perkins, Rosemary 110

permaculture 267, 268–9, 270–71
pesticides/insecticides 3, 5, 53, 65, 66,
    74, 80, 82, 85–6, 87–8, *88*, 90, 108,
    112, 114–18, 142, 167, 189, 190,
    202–3, 220, 221, 246–8, 258,
    260–62, 264, 273, 274, 276, 284,
    291–2, 295–6
    aldrin 87
    avermectins 87, *88*
    Bacillus thuringiensis sprays 87
    chlorpyrifos 87
    clothianidin 103, 104, 111
    DDT 86–7, *88*, 89, 90, 91, 99, 111, 115,
    184–5
    dicarboxamide fungicides 87
    dieldrin 87, 184–5
    fipronil 87, 110, 115
    'gene-silencing' 198–9
    imidacloprid 90, 92–4, *96*, 98, *99*, 103,
    104, 105, 110, 111
    imidazoles 87
    ivermectins *116*, 117
    malathion 87
    neonicotinoids 23–4, 87, 90–98, *99*,
    99–100, 101–2, 103–6, *106*, *107*,
    107–11, 115, 125, 202, 247, 292
    organochlorides 87, 89
    organophosphates 87, 90
    parathion 87
    phosmet 87
    pyrimidines 87
    spinosad 87
    thiamethoxam 103, 104, 111
    triazoles 87 *see also* fungicides; herbicides
petrol fumes 191
phosphates 136–9, 140, 214
phosphorus 136
photosynthesis 136
Pilliod, Alberta *and* Alva 132
plants
    exotic 181
    extinct 32, 124–5
    invasive 26, 181, 186–7, 280
    pollinator-friendly 245–6, 254, 289, 293
    see also grasses; weeds; wildflowers
plastics 86, 247, 248, 253
    and pollution 41, 190, 213–14, 218, 231–2,
    242, 247
Plaw Hatch biodynamic farm, West Sussex
    269–71
PloS ONE (journal) 51
Plymouth, University of 141
politicians 229, 233–4, 235–6, 289, 291–3,
    295–7, 299–300
Pollard, Ernie 55
pollination/pollinators 3, 4, 5, 24–5, 26, 30
    and n, 31–2
    'buzz' 149
    by hand 31, 211–12
    by orchid bees 84

and pollinator-friendly plants 245–6, 254, 289, 293
pollutants/pollution 216, 217–18, 229, 230, 254
  air 191–2, 216, 254
  chemicals 190, *see* fertilisers; herbicides; pesticides
  and geoengineering 192–4, 218
  heavy metals 189–90, 193, 218
  light 174–7, 201, 293
  particulates 162, 191–2
  petrol fumes 191
  plastics 41, 190, 213–14, 218, 231–2, 242, 247
  smog and smoke 162, 219
  of soil 89, 92, 98–9, 99, 100–3, 106, 111, 117, 126, 221, 259
  of water 3, 22, 23–4, 77, 79, 80 *and n*, 98, 105–7, 110, 117, 142–3, 144, 206, 224, 259, 272, 273, 276, 292–3
polybrominated diphenyl ethers (PBDEs) 190
polychlorinated biphenyls (PCBs) 89, 190
ponds 65, 79, 126, 207, 237, 240, 248, 249, 250, 254, 289, 290, 293
pond skaters 248, 290
population, human 174, 217, 258
potash 139, 140
potassium 136, 139
Powney, Gary 59, *60*, 65
prickly pears 26
processed food 259, 263, 274
pterosaurs 12
Purvis, Professor Andy 279
Putin, Vladimir 172

Quammen, David: *Song of the Dodo* 82

rabbits 48, 148, 179–80, 215
radiofrequency radiation 194–5
railway cuttings/embankments 4, 244, 254
rainfall 160, 162
rainforests, tropical 75–6, 83, 160, 162, 169, 216, 220, 281–2
Ramsar Convention sites 278–9
rats 48, 179, 183
reptiles 21, 48
  slow worms 34, 213 *see also* lizards
'resilience' 205–6
rewilding 5, 280, 282, 285, 291, 297
Rio de Janeiro, Brazil 168, 219
  Rio Convention on Biological Diversity 172
rivers 2, 143, 162, 221, 285
  pollution 3, 58, 77, 80, 89, 99, 105, 110, 113, 118, 142, 143, 272, 281, 293
road verges 4, 129, 190, 191, 203, 231–2, 244, 247, 249, 254, 284, 294
Roszak, Theodore 251
Rothamsted, Harpenden 138–9
  Rothamsted Research 57, 139*n*

roundabouts 4, 190, 244, 249, 250, 254, 284, 294
Roundup (herbicide) 114*n*, 120, *121*, 129, 131
  'Roundup-ready' crops 121–2, 198
Royal Horticultural Society 245
Royal Society for the Protection of Birds (RSPB) 278, 279
Rumsfeld, Donald 189, 199
Rundlöf, Maj *95*, 96
Russia 162, 171–2, 185
rusts (disease) 112, 181
Rutherford, Daniel 140

Sahara Desert 160, 220
Salisbury Plain 79
Sánchez-Bayo, Francisco, and Wyckhuys, Kris: studies 63–4
Saskatchewan, University of 105
Saudi Arabia 171
Scandinavia 164, 219 *see also* Denmark; Norway; Sweden
Schieffelin, Eugene 179
schools 70, 237–41, 249, 253*n*, 254, 288
Schrader, Gerhard 87, 90
Schulze, Svenja 235–6
Science (journal) 51, 94
Scotland 34, 58, 77, 92, 101, 164, 249, 282, 284
sea levels, rising 160–61, 162
Seibold, Sebastian 52–3
Shadow (insecticide) 114*n*
Shanghai, China 160, 219
Sharlow, Matt 103*n*
'shifting-baseline syndrome' 68
Shortall, Chris 57
Siberia 162
silkworms 21–2
silverfish 12, 19, 117, 118
'sinks' 81, 176
Sites of Special Scientific Interest (SSSIs) 279, 285
slow worms 34, 213
slug pellets 80, 142
slugs 34, 35, 70, 213
Sluijs, Jeroen van der 105
snails 165
  land snails 180
  rosy wolfsnails 180–81
  tree snails 181
social media 289
Söder, Markus 235
soil(s) 3, 4, 5, 10, 26, 27, 29, 89, 136, 159, 215, 240, 262, 264, 265, 266–7, 271, 274, 280, 282, 284–5,
  erosion 76, 80 *and n*, 206, 216, 217, 220, 221, 229, 266–7, 268, 276, 281
  fertilising *see* fertilisers
  pollution 89, 92, 98–9, *99*, 100–3, *106*, 111, 117, 126, 221, 259 *see also* peatlands

Somerset Levels 78
South America 63
  bee disease 154
  bees 60, 84, 149
  guano 137
  insect eating 21
  invasive species from 26, 28, 180, 181,
    183, 186
  leafcutter ants 72
  sloth moths 38 *see also* Chile; Brazil;
    Colombia
soya bean production 76
Spain 26, 77, 148, 162, 175, 196
Special Areas of Conservation 279
spiders 10, 12, 21, 119, 175, 186
  wolf spiders 286
springtails 26–7
squashes, pollinating 212
Steiner, Rudolf 269
stick insects 40
Stirling, Scotland 249
  University 92–4
storms 161, 162, 166
Strepsiptera 40
Sumatra 162
Sussex University 100, 111, 115, 123
Sweden 95, 135, 162
Syngenta (company) 92, 95–6, 133–4

Tasmania 149, 186, 268
taxes, agrochemical 273, 295–6
taxonomy 300
telephone masts 194–5
termites 16, 23*n*, 25*n*, 156
  Neocaptitermes taracue 156
  queen 15*n*, 16
Texas: universities 125, 186
Thames Estuary 79
Thunberg, Greta 231
ticks 10
toads *see* amphibians
towns *see* cities
traffic 190, 191
tree hoppers 37–8, *39*
  ball-bearing 38
trees 219, 229*n*, 254, 267–8, 300
  ash 185–6
  bottlebrush 181
  box 181
  Caledonian pine 282
  chestnut 214–15, 294
  elm 184–5
  flowering 248, 294
  fruit 182, 290–1, 294
  gum 181
  hawthorn 294
  Hirtella physophora 277
  lime 294
  oak 165, 214
  rowan 294

  tea 181
  wayfaring 294
Trenberth, Kevin 194
trinitrocellulose (TNT) 140
Trump, Donald 157, 171, 236
trypanosomes 146
Tuncak, Baskut 134

United Nations Intergovernmental Panel on
  Climate Change 194
United States of America
  bees 30*n*, 60, 91, 112, 147, 148, 149, 151,
    152, 153, 163
  birds 61, 179
  'community gardens' *see* allotments
  deforestation 139
  Endangered Species Act 299
  Environmental Protection Agency (EPA)
    127, 128, 129, 130, 236
  dengue fever 168
  insect decline 69, 70
  invasive species 179, 181–2, 184, 185
  monarch butterfly 3, 123, 173, 203–4
  National Parks 280, 299
  neocotinoids/pesticides 91–2, 108, 117
  periodical cicadas 226
  US Fish and Wildlife Service (FWS) 299
  wheat belt 167, 220
urban areas *see* cities/towns
Utrecht, University of 105

Varroa destructor mites 91, 151–2, 153, 154
veganism/vegetarianism 241, 263*n*
vegetables, growing 265–6, 290–1, 293
Victoria, Queen 34
viruses, insect 146, 147
  deformed wing virus 97, 152–3, 154

Wagstaff, Aimee 131
Warren, Martin 164
wasps 14, 16, 21, 22, 25, 26, 33, 60, 61, 89,
  109, 152, 284
  Cotesia glomerata 26, 256
  emerald cockroach wasps 67
  fairy wasps 25*n*, 256
  Lysibia nana 256
  paper wasps 26
  parasitoid wasps 155
water 229
  drinking water 80, 113, 143, 224, 273
  groundwater 82, 98, 113, 216 *see also*
    droughts; freshwater habitats; lakes;
    ponds; rivers
weeds 120, 121, 122–3, 124, 246, 290, 291, 292
weevils 38
wetas, giant 35–6, 183
wetlands 278–9
White, Gilbert 278
whiteflies 116, 213, 291
Whiteside, Mark 193

wildfires 159, *161*, 162, 163, 166, 191–2, 219, 222
wildflowers and plants 2, 4, *5*, 24, 31–2, 77, 78, 80, 97, 99–103, 104, *106*, 141, 190–91, 245, 248, 249–50, 254, 284, 289–90, 294
　bird's foot trefoil 245, 246, 289
　buttercups 246
　clover 78, 149–50, 246
　corn cockles 122, 124
　cornflowers 24, 122, 124
　corn marigolds 122, 124
　cowslips 69, 141
　daisies 246
　dandelions 97, 99, 124, 186, 246, 290
　docks 141
　forget-me-nots 24, 101
　foxgloves 24, 245
　greater butterfly orchids 79
　herb Robert 290
　Himalayan balsam 181
　hogweed 101, 141, 181, 290
　horseshoe vetch 167
　ivy 289
　Japanese knotweed 181
　kidney vetch 78
　knapweed 78, 186
　lady's smock 245, 289
　milkweed 123
　nettles 121, 141, 212, 289
　orchids 84
　ox-eye daisies 186
　poppies 24, 101, 122, 124
　'Pyrethrum daisy' (*Chrysanthemum cinerariaefolium*) 85
　ragwort 290
　red bartsia 78
　St John's wort 101
　selfheal 246
　thistles 101, 122 *and n*, 186
　violets 101
　viper's bugloss 78, 186, 245
　white deadnettle 245 *see also* crops; grasses; plants
Wildlife Trusts 278, 279
Wilson, E. O. 16, 19, 29–30, 75,
　Biophilia 251
　Half-Earth 281
　The Theory of Island Biogeography (with MacArthur) 281
'windshield phenomenon' 69
Winston, Lord Robert 20, 36
wolves 284, 285
wood piles 291
Woodcock, Ben 95, 96
Woodlands Trust 279
woodlice 11, 117–18, 165, 291
World Health Organization 127, 196
'World Scientists' Warning to Humanity' (2017) 230
World Wildlife Fund 48
worms 291
　bagworms *see* moths
　earthworms 240
　maguey worms *see* caterpillars
　mopanie worms *see* caterpillars
　nematode worms 146
　silkworms *see* caterpillars
　slow worms *see* reptiles
　velvet worms 14
Wright, Geri 102*n*
Wyckhuys, Kris *see* Sánchez-Bayo, Francisco

Xerces Society 54

Zhang Luoping 131–2
zooplankton 107, *107*